Studies in Systems, Decision and Control

Volume 141

Series editor

Janusz Kacprzyk, Polish Academy of Sciences, Warsaw, Poland
e-mail: kacprzyk@ibspan.waw.pl

The series "Studies in Systems, Decision and Control" (SSDC) covers both new developments and advances, as well as the state of the art, in the various areas of broadly perceived systems, decision making and control- quickly, up to date and with a high quality. The intent is to cover the theory, applications, and perspectives on the state of the art and future developments relevant to systems, decision making, control, complex processes and related areas, as embedded in the fields of engineering, computer science, physics, economics, social and life sciences, as well as the paradigms and methodologies behind them. The series contains monographs, textbooks, lecture notes and edited volumes in systems, decision making and control spanning the areas of Cyber-Physical Systems, Autonomous Systems, Sensor Networks, Control Systems, Energy Systems, Automotive Systems, Biological Systems, Vehicular Networking and Connected Vehicles, Aerospace Systems, Automation, Manufacturing, Smart Grids, Nonlinear Systems, Power Systems, Robotics, Social Systems, Economic Systems and other. Of particular value to both the contributors and the readership are the short publication timeframe and the world-wide distribution and exposure which enable both a wide and rapid dissemination of research output.

More information about this series at http://www.springer.com/series/13304

Rolf Dornberger
Editor

Business Information Systems and Technology 4.0

New Trends in the Age of Digital Change

 Springer

Editor
Rolf Dornberger
Institute for Information Systems
University of Applied Sciences and Arts
  Northwestern Switzerland
Basel
Switzerland

ISSN 2198-4182   ISSN 2198-4190   (electronic)
Studies in Systems, Decision and Control
ISBN 978-3-030-08963-4   ISBN 978-3-319-74322-6   (eBook)
https://doi.org/10.1007/978-3-319-74322-6

This Springer imprint is published by Springer Nature
The registered company is Springer International Publishing AG
The registered company address is: Gewerbestrasse 11, 6330 Cham, Switzerland

Preface

It is said that we live in the digitalization age, or even in the age of digital change and digital transformation. What does this mean and where is it leading us?

These and further questions occupy our minds today—in our business and private lives as well as in our research. It starts with the word "digitalization", which, as often mentioned, originates from the word "digitization". However, whereas digitization outlines the process of transforming data and information from an analog format into its digital counterpart, digitalization describes how the worlds of business and society are changing due to the opportunities offered by digitization. The world itself, however, will not become digital: Not everything will be zero or one, black or white, head or tail; moreover, we are not heading for a binary future. Instead, we will have an increasing number of polymorphic solutions between zero and one, more grayscale and colorful solutions that will shape our future more intensively than ever before in human history.

This book has been written by a group of professors, lecturers, and researchers (including external colleagues) working in the field of Information Systems, Business Informatics, Computer Science, Business Administration, and Management at the School of Business of the University of Applied Sciences and Arts Northwestern Switzerland. As a university of applied sciences, our mission is to link academia and practical experience within our research and teaching activities. We focus our applied research and development on answering questions that arise from practice, and we transfer the research results to practical application. Thus, the style of the chapters of this book follows the philosophy of applied sciences by balancing the degree of profundity and rigor in our research with its translation into relevance in practice.

As the Institute for Information Systems is located at a business school, our original focus was on bridging the gap between business and IT, as well as aligning business and IT in the context of organizations. We have now broadened our focus to researching the impact of new IT-related technologies and IT-supported methods—including hot topics such as agile process management, artificial intelligence, robotics, management of complex systems, cybersecurity, machine ethics, digital business transformation, etc. Furthermore, we are designing such smart systems, evaluating their interaction with humans and investigating their use in business and society.

This finally led to the idea of writing a book about our passion. We modestly named it "Business Information Systems and Technology 4.0—New Trends in the Age of Digital Change". In this book, we present the trends that we consider important in the age of digital change and their concrete use to boost the innovativeness and efficiency of organizations, discover new business opportunities, and, lastly, manage digital transformation in order to address the challenges within ever-changing business and societal environments. The first chapter is the key chapter, presenting our model of "Digitalization: Yesterday, today and tomorrow". It introduces the topics presented in this book and links them into one overarching theme.

Through the combination of application orientation and research depth, this book will be of interest and value to all who intend to leverage these trends to take their business beyond today's possibilities. As we contribute to the existing body of knowledge in the specific domains, this book will also be of interest and value to researchers. The future is not yet known, but it is waiting to be discovered by you, our readers.

As the editor of the book, I would like to express my gratitude to our employer, the University of Applied Sciences and Arts Northwestern Switzerland, and would particularly like to thank the School of Business for supporting the book by granting additional hours to the authors to write their chapters. Furthermore, my thanks go to all our authors, who made an excellent contribution to this work, providing insights into a variety of research fields, valuable research findings, and outstanding teaching content. Moreover, my special thanks go to Prof. Dr. Thomas Hanne and Prof. Dr. Uwe Leimstoll, who scientifically, thoroughly, and independently reviewed all the contributions, giving valuable, constructive feedback, and not hesitating to challenge us to make qualitative improvements to the content. Further, I would like to thank Vivienne Jia Zhong, who efficiently and competently coordinated the contributions of more than 30 authors, compiled in 18 chapters, and oversaw the progress of the book. My thanks also go to Christine Lorgé and Margaret Oertig, who carried out the proofreading of this book with diligence and commitment to language clarity. In addition, I would like to thank all our families for their enduring patience and great support.

Basel, Switzerland   Prof. Dr. Rolf Dornberger
October 2017

Contents

Digitalization: Yesterday, Today and Tomorrow

Rolf Dornberger, Terry Inglese, Safak Korkut and Vivienne Jia Zhong

Abstract The rapid development of digital technologies is making organizations rethink their business models and processes. This is resulting in a massive digital transformation of the economy and society. New trends are emerging at a fast pace and some of them might vanish soon. In order to investigate the opportunities and challenges behind these trends and make a sound prediction of their further development, it is necessary to understand the evolution of information and communication technology, which was and still is intended to provide support in the management of personal and business tasks. For this purpose, we present our model "Digitalization: Yesterday, today and tomorrow", which provides a brief summary of the development and rise of computational technology, resulting in changes in the interaction both of humans with computers and between humans and computers, and shows how individuals, business and the government have been adapting to these changes. We identified four streams of development: Early Information Systems, the E-Business Applications, the Web 2.0 Revolution and the renaissance of the Artificial Intelligence; a fifth stream remains unnamed, as we do not yet know where these fast-paced developments will lead us.

Keywords Digitalization · Information system · E-Business · Web 2.0
Artificial intelligence

R. Dornberger (✉) · T. Inglese · S. Korkut · V. J. Zhong
Institute for Information Systems, University of Applied Sciences and Arts Northwestern
Switzerland, Peter Merian-Strasse 86, 4002 Basel, Switzerland
e-mail: rolf.dornberger@fhnw.ch

T. Inglese
e-mail: terry.inglese@fhnw.ch

S. Korkut
e-mail: safak.korkut@fhnw.ch

V. J. Zhong
e-mail: viviennejia.zhong@fhnw.ch

© Springer International Publishing AG 2018
R. Dornberger (ed.), *Business Information Systems and Technology 4.0*,
Studies in Systems, Decision and Control 141,
https://doi.org/10.1007/978-3-319-74322-6_1

1   Introduction

The world is changing, probably faster today than ever before in human history. If we use a pattern to visualize this statement figuratively, the change might be described as follows: In early human imagination, the world was considered to be flat and assumed to be a kind of two-dimensional disk, where you can fall off its edge. Around 500 B.C., the Greek philosopher Aristotle spoke of the world as being a globe, the diameter of which Eratosthenes (approx. two centuries later) calculated very roughly and identified the earth as a three-dimensional shape without edges. However, it was not until 2,000 years later that Fernão de Magalhães was able to experimentally prove its globe shape by circumnavigating the earth at the beginning of the 16th century. Five hundred years later, in the course of globalization, Thomas Friedman declared that "The world is flat" (Friedman 2005), thus returning to the idea of an economically interconnected, but flat, two-dimensional world. Moreover, it is predicted today that "we live in exponential times" (Demirdjian 2015), suggesting that our world has shrunk to a one-dimensional exponential curve, where key issues are expanding at exponential speed and growth.

The pattern from a geographically 2D flat world via a 3D round globe back to an interconnected world economy and society (flat = 2D) ends up in an exponential 1D curve. Furthermore, the futurist Ray Kurzweil predicts that "the singularity is near" (Kurzweil 2005; Galeon and Reedy 2017) meaning that the point in time where machines are smarter than human beings thanks to advances in technology will soon be reached: Is the world shrinking to zero dimensions and vanishing?

We, my colleagues and I, are involved in teaching and research on such related topics, where digitalization aims to change the foundations, the mindset and the thinking about how people live and work—together and/or alone—in a world changing faster than ever before. New technologies, opportunities, business models and threats are emerging at a tremendous speed, only to vanish again as quickly as they appeared. To investigate this and to understand what is happening today and tomorrow, we have to understand the development of how information and communication technology (ICT), e.g. computers, software, the Internet etc., was intended to support the management of personal and business tasks.

To provide a common umbrella for the discussions in the following book chapters, we refer to our model "Digitalization: Yesterday, today and tomorrow" (see Fig. 1). This model provides a short summary of the development and rise of computational technology over the last decades, resulting in changes in the interaction both of humans with computers and between humans and computers, and shows how individuals, business and the government have been adapting to these changes. As depicted in Fig. 1, we identified that the literature about information systems generally proposes four streams of development: (early) Information Systems, E-Business Applications, Web 2.0 Revolution and (the renaissance of) Artificial Intelligence. Finally, but importantly, a fifth development stream is still written with a question mark, because we do not know where all this rapid development will lead us to and

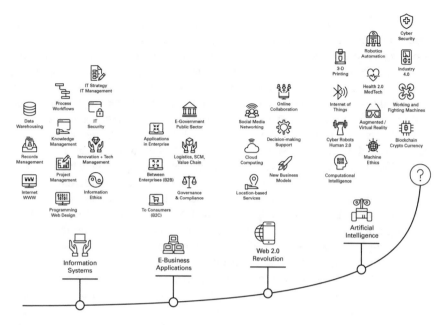

Fig. 1 Digitalization: Yesterday, today and tomorrow

what to call it. The question mark opens up new questions, which we—like everyone else—would love to answer as soon as possible.

In the following sections, we describe each of these streams. In particular, we elaborate the Artificial Intelligence stream in more detail, because the recent innovations we have been witnessing are heavily affected by it.

2 Information Systems

The first development stream is about the establishment of early information systems, which mainly affects corporate organizations. In the second half of the last century, computers were increasingly found to be useful in companies. An irony of history is that, in 1943, the former chairman and CEO of IBM, Thomas J. Watson, is claimed to have said that "there is a world market for maybe five computers" (Wikimedia Foundation, Inc. 2017). IBM went on to make a multi-billion dollar business out of it and now owns the most "intelligent" computer, called WATSON. Later on, in the 1970s, computers multiplied at the workplace, and with time were referred to as personal computers (PCs), thus indicating that everyone should have one. Computers attracted humans for private purposes, too, e.g. for writing letters and doing calculations, but mostly for gaming.

In the 1980s, computers were mainly used to enter, store, process and print data (data in the real sense of numbers and characters). Thanks to research in the field of information and knowledge management, we learned how to create and develop information and knowledge. The more computers there were on the market and at home, the more data were generated and processed and the more information and knowledge were piled up—theoretically. However, beforehand, data warehouses were needed to store all these data in databases, where, for example, records management is a small subtopic of a particular data handling issue in business. Process management was developed to manage the data flow effectively and to permit the sequencing of repetitive tasks using the data. A better understanding of workflow management ensures that the correct artefacts are found and processes are automatized.

Improvements to hardware and software, but also in project management, programming, the connection of computers to the Internet and to the World Wide Web (WWW), and the design of user interfaces brought us a new form of information systems, starting in the early 1990s. Furthermore, to integrate these powerful information systems into the context of work, companies needed a well-elaborated information technology (IT) strategy and strong IT management: IT became strategic. In order to bring in new ideas and technologies, innovation and technology management also became important. In parallel, IT security started to become a major issue to protect data and information—and to protect users and unaware, uninvolved people. Overall, information ethics was developed to assure the ethical use of information systems and the underlying technologies, data and processes.

3 E-Business Applications

Starting in the late 1990s, the business sector and the government developed their own usage of information systems. The concept of E-business emerged and attention was drawn to the development of IT applications that supported this concept. Companies then started developing business software such as enterprise resource planning systems and other enterprise applications, for the purpose of working more smoothly together, either internally within the organization, or interconnected with other enterprises (known as B2B, business-to-business), or with consumers (known as B2C, business-to-consumer). The intention was to promote the collaboration of enterprises with suppliers, customers and other organizations by managing and streamlining the data flow via electronically-supported processes. The idea was—to give an example—to extend classical logistics to supply chain management (SCM) and, even further, to overall value chain management.

In addition, governments from state level down to community level identified the potentialities of the IT, especially for providing citizens with a platform for all sorts of civic engagement, such as elections, electronic tax declaration systems and other types of civic participation. This became known as e-government.

In order to handle all these developments, the discipline of governance and compliance began its ascent, upgrading the legal issues and the so-called information ethics concept to standards of good and sensible conduct.

4 The Web 2.0 Revolution

In the early years of the 2000s, two new game-players made their appearance: the smartphone (first the iPhone in 2007 (The Telegraph 2017)) and social media (Kaplan and Haenlein 2010).

With the advent of smartphones came a major change in the instant availability of computing devices: It was now possible to carry our computers and our access to the Internet in our pockets; our computers were no longer sitting on our desks or stored in our bags. This change made new business models possible by applying new mobile commerce scenarios, such as mobile banking, online shopping, speed dating and so on. The scaling up of the business model to millions of users has led us to refer to platform capitalism (Lobo 2014), where globally accepted information ethics principles are urgently needed. Furthermore, constant access to the Internet provides organizations with the possibility to track our location to offer location-based services.

Moreover, with the Web 2.0, we are now not only consuming what is offered through the Internet; we are also simultaneously offering information to the Internet through a sort of non-negotiated and spontaneous online collaboration between all kinds of combinations of individuals, companies and organizations.

To facilitate collaboration, cloud computing was established as a solution. We neither need to concern ourselves with the physical storage of our data nor with the hosting of software required for running a business. Whatever we need, we can access it immediately from the Internet. We also do not need to bother about software updates. For instance, our mobile device is continuously being improved and enhanced in the background, and thus provides decision-making support for all kinds of situations (e.g. shopping suggestions, dining recommendations and leisure tips).

Social media provides the ultimate convergence of all types of networking, communication and collaboration in the digital scenarios already highlighted. Online users are becoming the producers and consumers of new types of content. Whether we want it or not, they are provoking new, rapid, social and societal transformations, through the interconnectedness of online networks.

5 Artificial Intelligence—The Renaissance of AI

Research in Artificial Intelligence (AI) already started in the second half of the last century, but led to disillusionment in the 1990s, because the theory predicted more than computational power was able to prove. In the meantime, computational power is constantly increasing—still bravely following Moore's law (Denning and Lewis 2016)—speeding up algorithms and entire computer applications at an unknown pace. Thus, data can be quickly transferred, algorithms requiring immense computational power can be widely applied, information and knowledge can be quickly processed (mined) from any kind of data, and machines are starting to learn. This is the renaissance of AI! Below we comment on the meaningful developments within this stream.

Within this epoch, an important key concept is the Internet of Things (IoT), where everything is connected with everything. Every device, which needs electronic power will be linked to the Internet and—theoretically—the Internet will "know" about every device in the world connected to it. Moreover, people are connected too with their mobile devices. Consequently, the IoT is becoming the driving force for sensing data and information of all kinds, because more and more machines are being equipped with a vast number of sensors. They report all sorts of data to the Internet, e.g. acceleration, rotation, location, Wi-Fi connectivity, noise as well as spoken words, light intensity as well as camera pictures, and temperatures as well as sweat levels. For example, Industry 4.0 has the vision to connect all production machinery to the Internet (Federal Ministry of Education and Research n.d.). New value-added services, for instance, predictive maintenance, are just at the beginning. However, the constant connection of everything to the Internet is prone to cyber-attacks. Consequently, cyber security technologies play a more important role in sustaining business today than ever before.

Another important game changer has come on the scene, extending the possibilities of IoT and Industry 4.0 to robotics automation. In the past, computing machines were mainly stationary or, in the best case, carried around in our pockets like our smartphones. However, with the rise of mobile robots, they are now able to approach us—whether we want this or not. This phenomenon is known as the rise of the autonomous (Tan and Ng 2017; LaPlante 2017), where machines now approach humans—instead of the other way about. For example, self-driving cars identify their passengers on the pavement and pick them up (Hawkins 2017), pizza delivery drones bring food directly to the fifteenth floor of a skyscraper, knocking on the kitchen window (Reid 2016), police robots serve justice (Cellan-Jones 2017), and healthcare robots reassure us that they will take care of everything (Robinson et al. 2014). However, we do not have to worry at all. Such working machines are still our utopia, although turning working machines into fighting machines is our dystopia.

Another remarkable technology, 3D printing is emerging, with the potential to change the entire supply chain, because it opens up new possibilities for production by reshaping the distribution processes of goods. Within 3D printing scenarios, logistics processes are becoming local, because we are able to produce and print products instantly and locally, on the spot, where we want to receive them, for example, a nice T-shirt just after we get up, a delicious hamburger for lunch, a new leg after a severe car accident, etc.

Within a Health 2.0 scenario, we might use Medtech to print an artificial heart after a heart attack (Cohrs et al. 2017). Furthermore, a printed body part may lead to human augmentation (Mann et al. 2016), offering the opportunity of using, for example, a third arm while doing all sorts of errands (e.g. two arms hold the heavy parcel, and the hand on the third arm rings the doorbell). Alternatively, we may print or actually breed our cyber robots. Are they already Human 2.0? Alternatively, are we the "Humans 2.0" and they go even beyond that?

Everything (and everyone) is becoming "cyber". The differences between the real and virtual world are diminishing: Augmented reality is enhancing our visual perception of the world by projecting fictive things in our field of vision. Virtual reality allows us to completely immerse ourselves into a virtual world. The role of cash as a payment method is decreasing, not even a credit card is hip anymore. The new trend is cryptocurrencies (Swan 2015; Luther 2016), where neither governmental institutions nor banks guarantee their value anymore, instead a new kind of trust information is stored in so-called blockchains (Swan 2015), decentralized somewhere in the Internet cloud.

Computing machines are becoming increasingly intelligent. This artificial machine intelligence is probably still different to human intelligence—however we define intelligence. Nevertheless, according to Nick Bostrom, machine intelligence is so powerful that it will outperform the human intelligence of all human brains already within this century (Bostrom 2014). Bostrom calls this effect superintelligence. The underlying methods and algorithms belong to artificial intelligence, perhaps to the field of computational intelligence, which uses nature-inspired methods to solve complex real-world problems (Hanne and Dornberger 2017).

We now return to the question mark in Fig. 1. We are equipping machines with an increasing number of AI methods and are still raising computational power, letting them connect via the IoT, granting them autonomous mobility, and thus making them more intelligent than us. What consequences superintelligence will have is completely unknown to us. As humans, we are good at creativity, empathy, designing, composing music and so on. We do not know when computers will achieve this kind of intelligence, but, based on Nick Bostrom, Ray Kurzweil and other thought leaders, we are quite sure that they will achieve superintelligence within the next few decades (Bostrom 2014; Kurzweil 2005; Galeon and Reedy 2017). What will happen then? We should definitely try to introduce machine ethics to computers and robots. However, will this work and will it be enough? The following example illustrates this concern: On August 21st, 2017, 113 leaders and CEOs of companies in robotics and informatics from all over the world presented a jointly signed letter to the United Nation commission to ask for the immediate regulation of "certain conventional weapons" without any kind of doubt about the potential threats in repurposing artificial intelligence and robotics from a societal benefit towards a third revolution in warfare. *"Once this Pandora's box is opened, it will be hard to close. We therefore implore the High Contracting Parties to find a way to protect us all from these dangers"* (University of New South Wales 2017; Future of Life Institute 2017). It seems that the request of these leaders and CEOs to the UNO commission should be discussed collectively and it should be addressed to all of us, where we can be part of a collective decision, which rewards "common sense" and respect for all sorts of creatures, humans as well as nonhumans.

The question mark at the end of the exponential curve of our model "Digitalization: Yesterday, today and tomorrow" (in Fig. 1) is perhaps an appropriate answer to the question about "tomorrow". We have it in our hands, today, to shape our "tomorrow" collectively and responsibly.

6 Organization of this Book

This book, named "Business Information Systems and Technology 4.0—New Trends in the Age of Digital Change", reflects on information systems, new technologies (which we summarize under the umbrella term Technology 4.0 in imitation of the term Industry 4.0) and on how digitalization is bringing about a change in business and society. The chapters are organized according to four meta-topics based on the sequence of Fig. 1 "Digitalization: Yesterday, today and tomorrow":

(a) Information Systems
(b) E-Business Applications
(c) Web 2.0 Revolution (and preliminary ending with)
(d) Artificial Intelligence.

Overall, the chapters address research topics such as digital transformation management, e-business, knowledge representation, the impact of digitalization on higher education, human computer interaction and computational intelligence. Finally, the book chapters are organized as follows:

(a) Information Systems

- "ERP Systems towards Digital Transformation": The new role of enterprise resource planning systems as a consequent extension of information systems.
- "Determining Information Relevance based on Personalization Techniques to Meet Specific User Needs": New solutions for information and knowledge management.
- "Case-based Reasoning for Process Experience": A new approach to integrated case-based reasoning for learning and experience management processes.
- "Road to Agile Requirements Engineering: Lessons Learned from a Web App Project": An explanation and best practices from agile requirements engineering embedded in project management.

(b) E-Business Applications

- "E-Business in the Era of Digital Transformation": The development of e-business over the years and its continuing great importance for understanding digitalization.
- "Digitalizing B2B Business Processes—The Learnings From E-Invoicing": Best practices of digitalizing complex business-to-business (B2B) processes based on the example of e-invoicing.

- "Marketing Automation: A Project Framework in Support of Digital Transformation": An overview of possibilities in marketing automation ranging from strategic considerations, customer journey analysis to use cases along with data management, content marketing and channel management.

(c) Web 2.0 Revolution

- "FHNW Maturity Models for Cloud and Enterprise IT": A new model to assess the digitalization readiness of enterprises and proposing the consequent next steps.
- "Digital Transformation Management and Digital Business Development": A new management concept for companies adapting their business to ongoing digital transformation.
- "Using Feedback Systems Thinking to Explore Theories of Digital Business for Medtech Companies": Supporting decision-making by applying feedback systems thinking to the example of Medtech companies.
- "Ontology-Based Metamodeling": A new method for decision making based on the same internal knowledge representation for humans and machines.

(d) Artificial Intelligence

- "Searching and Browsing in Historical Documents—State of the Art and Novel Approaches for Template-Based Keyword Spotting": Different state-of-the-art and novel approaches for keyword spotting (online, offline, without any a priori learning of a model, etc.) for the digitalization of handwritten documents.
- "How to Teach Blockchain in a Business School": The content and the preparation of teaching curricula related to the importance-gaining topic of blockchain.
- "Computational Intelligence in Modelling, Simulation, Optimization, and Control": Computational intelligence comprising nature-inspired AI methods with a focus on modelling, simulation and optimization in logistics and robotics.
- "Innovation Potential for Human Computer Interaction Domains in the Digital Enterprise": Innovative application scenarios of human computer interaction applying touch screen and natural language processing.
- "Prototype-Based Research on Immersive Virtual Reality and on Self-Replicating Robots": Summary and discussion of research topics in the field of virtual reality and self-replicating robots using gadgets and prototype-based research.
- "Co-Robots from an Ethical Perspective": Information and technology ethics, machine ethics and business ethics in the field of cooperation and collaboration robots.

Overall, the book offers a broad variety of topics on emerging trends in the age of digital change and illustrates concrete use cases for organizations to sustain their business in an ever-changing environment. We hope you will find our contributions inspiring.

References

Bostrom N (2014) Superintelligence: paths, dangers, strategies. Oxford University Press

Cellan-Jones R (2017) Dubai Police unveil robot officer. In: BBC News. http://www.bbc.com/news/technology-40026940. Accessed 11 Oct 2017

Cohrs NH, Petrou A, Loepfe M, Yliruka M, Schumacher CM, Kohll AX, Starck CT, Schmid Daners M, Meboldt M, Falk V, Stark WJ (2017) A soft total artificial heart—first concept evaluation on a hybrid mock circulation. Artif Organs 41:948–958. https://doi.org/10.1111/aor.12956

Demirdjian ZSA (2015) Challenges and opportunities in exponential times. Xlibris Corporation

Denning PJ, Lewis TG (2016) Exponential laws of computing growth. Commun ACM 60:54–65. https://doi.org/10.1145/2976758

Federal Ministry of Education and Research (n.d.) Industrie 4.0—BMBF. In: Bundesminist. Für Bild. Forsch.—BMBF. https://www.bmbf.de/de/zukunftsprojekt-industrie-4-0-848.html. Accessed 11 Oct 2017

Friedman TL (2005) The World is Flat: a brief history of the twenty-first century. Thorndike Press

Future of Life Institute (2017) Killer robots: World's top AI and robotics companies urge United Nations to ban lethal autonomous weapons. In: Future Life Institute. https://futureoflife.org/2017/08/20/killer-robots-worlds-top-ai-robotics-companies-urge-united-nations-ban-lethal-autonomous-weapons/. Accessed 23 Oct 2017

Galeon D, Reedy C (2017) Kurzweil claims that the singularity will happen by 2045. In: Futurism. https://futurism.com/kurzweil-claims-that-the-singularity-will-happen-by-2045/. Accessed 10 Oct 2017

Hanne T, Dornberger R (2017) Computational intelligence in logistics and supply chain management. Springer International Publishing, Cham

Hawkins AJ (2017) Uber's self-driving cars are now picking up passengers in Arizona. In: The Verge. https://www.theverge.com/2017/2/21/14687346/uber-self-driving-car-arizona-pilot-ducey-california. Accessed 11 Oct 2017

Kaplan AM, Haenlein M (2010) Users of the world, unite! The challenges and opportunities of social media. Bus Horiz 53:59–68. https://doi.org/10.1016/j.bushor.2009.09.003

Kurzweil R (2005) The singularity is near: when humans transcend biology. Viking

LaPlante A (2017) The rise of autonomous data platforms. In: OReilly media. https://www.oreilly.com/ideas/the-rise-of-autonomous-data-platforms. Accessed 16 Oct 2017

Lobo S (2014) S.P.O.N.—Die Mensch-Maschine: Auf dem Weg in die Dumpinghölle. In: Spieg. http://www.spiegel.de/netzwelt/netzpolitik/sascha-lobo-sharing-economy-wie-bei-uber-ist-plattform-kapitalismus-a-989584.html. Accessed 11 Oct 2017

Luther WJ (2016) Cryptocurrencies, network effects, and switching costs. Contemp Econ Policy 34:553–571. https://doi.org/10.1111/coep.12151

Mann S, Leonard B, Brin D, Serrano A, Ingle R, Nickerson K, Fisher C, Mathews S, Janzen R, Ali MA, Yang K, Scourboutakos P, Braverman D, Nerkar S, Malicki-Sanchez K, Harris ZP, Harris ZA, Damiani J, Button E (2016) Code of ethics on human augmentation: the three "Laws" I KurzweilAI. http://www.kurzweilai.net/code-of-ethics-on-human-augmentation-the-three-laws. Accessed 11 Oct 2017

Reid D (2016) Domino's delivers world's first ever pizza by drone. https://www.cnbc.com/2016/11/16/dominos-has-delivered-the-worlds-first-ever-pizza-by-drone-to-a-new-zealand-couple.html. Accessed 11 Oct 2017

Robinson H, MacDonald B, Broadbent E (2014) The role of healthcare robots for older people at home: a review. Int J Soc Robot 6:575–591. https://doi.org/10.1007/s12369-014-0242-2

Swan M (2015) Blockchain: blueprint for a new economy. O'Reilly Media, Inc

Tan X, Ng GH (2017) Why the rise of autonomous machines could help workers, according to robotics CEO. https://www.cnbc.com/2017/04/04/why-the-rise-of-autonomous-machines-could-help-workers-according-to-robotics-ceo.html. Accessed 16 Oct 2017

The Telegraph (2017) Apple's iPhone: a definitive history in pictures. In: The telegraph. http://www.telegraph.co.uk/technology/0/apples-iphone-definitive-history-pictures/. Accessed 10 Oct 2017

University of New South Wales (2017) An open letter to the United Nations convention on certain conventional weapons

Wikimedia Foundation, Inc. Thomas J. Watson (2017) In: Wikipedia. https://en.wikipedia.org/w/index.php?title=Thomas_J._Watson&oldid=803719944. Accessed 16 Oct 2017

Part I
Information Systems

ERP Systems Towards Digital Transformation

Petra Maria Asprion, Bettina Schneider and Frank Grimberg

Abstract Enterprise Resource Planning (ERP) systems employ highly integrated business software solutions that have existed for many years. Being the base of the IT application landscape of most enterprises, ERP systems remain fairly commoditized and scarcely leave room for differentiation. In view of the major digital transformations currently taking place, the role of ERP systems needs to be reconsidered. Geoffrey Moore's concept of "Systems of Engagement" stresses the need for enterprise applications to become more user-oriented in order to support collaboration and to empower employees. Based on this understanding, we developed a model that classifies how ERP systems can evolve depending on its people-centricity focus and its level of integration.

Keywords ERP systems · Systems of engagement · Digital transformation

1 Introduction

The introductory section discusses the evolution of digital transformation and potential implications for ERP systems. It outlines the relevance, purpose and objectives of the study as well as the methodology applied.

P. M. Asprion (✉) · B. Schneider · F. Grimberg
Institute for Information Systems, University of Applied Sciences and Arts Northwestern Switzerland, Peter Merian-Strasse 86, 4002 Basel, Switzerland
e-mail: petra.asprion@fhnw.ch

B. Schneider
e-mail: bettina.schneider@fhnw.ch

F. Grimberg
e-mail: frank.grimberg@fhnw.ch

© Springer International Publishing AG 2018
R. Dornberger (ed.), *Business Information Systems and Technology 4.0*,
Studies in Systems, Decision and Control 141,
https://doi.org/10.1007/978-3-319-74322-6_2

1.1 Digital Transformation and ERP Systems

The digital transformation of the economy and society can be seen as a revolution—with a similarly great impact as the Industrial Revolution of the nineteenth century. Originating in the 1960s with the introduction of the first computers, countless efforts were initiated to automate processes in nearly every industry, allowing for a further, accelerating step with the nascent internet and in particular the World Wide Web in the 1990s (Chalons and Dufft 2016). In short, digital transformation has been present ever since the 1960s and is very closely related to continuous technological innovations over time.

In the 1960s, with the release of mainframes, the management of huge amounts of data and automated calculations were made possible. In subsequent years, many other innovations such as client-server computing in the 1980s and ERP systems in the 1990s allowed for new productivity gains and significant transformations in the way enterprises and their employees work (Monk and Wagner 2013, p. 20 ff.).

During the 1990s and 2000s, the Third Industrial Revolution along with the dynamics of globalization transformed business and technology, manifesting itself via, for example, fully automated processes and web-based solutions (Lemke and Brenner 2015, p. 3 f.). During that period, the use of ERP systems expanded as evidenced by the rise of ERP vendors such as SAP, Oracle, JD Edwards, and others that permitted enterprises to automate and integrate complex processes (Mabert et al. 2001, p. 70; Kurbel 2013, p. 2). Since the beginning of the internet and other technical innovations, digitalization has no longer been limited to internal operations, but permits permanent and easy access through all supply chains and from all places with internet access (Venkatesh et al. 2012).

However, despite their incredible successes, it seems that ERP systems themselves are now in the age of digital transformation (as we will elaborate in Sect. 2.1) and that after a lifespan of more than 40 years, ERP systems have reached the end of their lifecycle. According to a study by Panorama Consulting (2016), there have been many indicators which have foreshadowed the "death" of ERP systems over the last decade. For example, current ERP software is available to enterprises of all sizes and industries with a wide variety of viable solutions. In respect of the operations, enterprises can decide between a multitude of options, e.g. web-based, cloud-based, on premise, mobile apps and a host of others (Bahssas et al. 2015). Nowadays, ERP systems are mature products that are vital for the companies' survival, but they offer hardly any opportunity to achieve competitive differentiation based on their use (Seddon 2005; Fosser et al. 2008; Moore 2011, p. 3).

Beyond this, ERP systems have a rather bad reputation. For example, despite their potential, many ERP implementations have failed or have not achieved the expected functionalities (Schwenk 2014, p. 42; Finger 2012, p. 7). Another issue is strong competitive pressure, leading to risk-averse enterprises with less tolerance for expensive and time-consuming ERP implementations (Stefanou 2014, p. 157).

For enterprises that want to lead their existing ERP system into the new digital age, it is recommended to define a smart strategic roadmap which considers (new) technology and concentrates on people and processes (Panorama Consulting 2016). In particular, the focus on people, and especially the user, seems attractive. Based on their original purpose, ERP systems strongly optimize data and processes, whereas soft-factors like user-experience and user-engagement have not been developed with the same intensity. Therefore, we raise this topic as our focus of interest in this study.

We assume that enterprises will generally need to decide about their future ERP landscape sooner rather than later. The management should answer the question: "do we only aim for (simple) ERP implementation/consolidation, or is it/should it be a part of a digital transformation initiative?" In the first case, the consequence is to automate the status quo; in the latter, the decision is to aspire to innovative disruptive technologies, implying potential changes to the current business model or even adding a new business model (Laudon and Laudon 2016, p. 119).

1.2 Purpose and Objectives

The purpose of this study is to accumulate knowledge for practical application but also for the academic discipline of information systems (IS). Specifically, the objectives of this study are to (1) point out the relation between digital transformation and enterprises' ERP systems, (2) stress the future role of collaboration using Moore's concept of Systems of Engagement, (3) enhance the existing theory focused on ERP transformation, and (4) guide future research by developing recommendations and putting forward a research agenda.

This study argues that enterprises using or planning to implement/consolidate an ERP system need to decide about the future role of (their) ERP system(s). More precisely, enterprises have to define whether they wish to realize a (simple) ERP implementation/consolidation, or a digital transformation. A certain focus in this respect is on collaboration and people, in particular on the new generation of so-called digital natives, who tend to bring different expectations to their work environment and the way they work (Prensky 2001a, b; Roberts 2005; Koutropoulos 2011; Moore 2011, p. 2).

1.3 Methodology

In analyzing our focus of interest, we use as primary method the recommendations of Tranfield et al. (2003) as well as Benbasat and Zmud (1999) to conduct systematic literature reviews. In general, the literature search aims to identify the sources focused on the research topic, and in addition, seeks to prove the relevance and rigor of the research (vom Brocke et al. 2009). Relevance is achieved by ensuring that already known aspects of the research are not investigated twice (Baker 2000).

In order to contribute to the discussions around the role of ERP systems in the digital age, we set a clear focus on the potential of collaborative aspects. For this reason, we base our study on the well-established concept of Systems of Engagement as described by Geoffrey Moore. This concept is referenced by both practitioners such as Forbes (Orosco 2015) or Hewlett Packard (Barkol et al. 2012) and academia, e.g. for a new IT consumerization theory (Niehaves et al. 2013; Köffer et al. 2014). We discuss selected aspects of Moore's concept to show the adaptability and relevance of his model in different business contexts (Sect. 2.3).

As a foundation, the Systems of Engagement concept will be used to design a straightforward model intended to support decision makers on ERP initiatives. We use a design science related approach. The approach supports our research objectives outlined above, with Hevner et al. (2004) providing seven guidelines for design science in IS, directing the research activities, especially regarding guideline 5 which addresses "research rigor".

Regarding our main method, we conducted a literature review with a search in the Web of Science (all journals) and in Google Scholar (top journals). Then, we enriched the matches from scientific literature by adapting practical knowledge from practitioner-oriented sources, for example the publications of Moore (2011, 2014). The search queries designed to select papers related to the topics "ERP systems", "ERP", "digital transformation", "digitalization", "digital business", "business age", "disruptive technologies" and "business technology".

The remainder of the paper is structured as follows. In the next section, we present a short explanation of the key concepts starting with disruptive technologies. Further, Moore's approach regarding Systems of Engagement and selected applications are introduced. In Sect. 3, the designed "ERP transformation model" is elaborated and in Sect. 4, we present a conclusion and an outlook of further research planned.

2 Key Concepts

In this section, the Systems of Engagement concept is described and aligned with ERP related perspectives. Based on exemplary cases the concept's broad field of potential applications is shown.

2.1 Disruptive Technologies

A wide discussion about the next disruptive industrial revolution started in recent years (Bauernhansl et al. 2014, V). In the German-speaking countries, the expression "Industry 4.0" (I4.0) emerged, following an initiative by the German government. High-income countries in particular need to exploit technological innovations in order to keep up with the competition and manage to stay ahead of the market. In order to achieve this goal, "cyber-physical systems" are required to merge the digital and the physical world (Sontow and Schürmeyer 2014, p. 19 f.). Related in this respect are the drivers of the Fourth Industrial Revolution, e.g. Internet of Things (IoT), big data or augmented reality (Kaufmann 2015, p. 5).

Many publications related to I4.0 start their journey with an evaluation of technological advances followed by a future prospective on how I4.0 will change the world of business. For example, Kaufmann (2015) sets the focus on how IoT affects existing or new business models. Sendler (2013) describes intelligent, connected systems, which are difficult to integrate and challenging for many industries and their products. Andelfinger and Hänisch (2017) discuss how cyber-physical systems will influence working environments. Botthof and Hartmann (2015) evaluate the future role of labor in a digitalized world.

With many more examples available, there is no doubt that technology and the new opportunities around digitalization will change the business world enormously. In addition, digital natives are entering the job market with different mindsets and expectations, not least regarding how they want to learn and work (Roberts 2005; Koutropoulos 2011). Enterprises that aim to attract this generation as customers and employees need to investigate the appropriate composition of their future enterprise IT setup (Moore 2011, p. 2). Thus, in the future, both technology and people have to be considered in an integrative way.

Moore analyzed digital disruptive technologies and their impact on executives and business life and published the results in a widely-known publication called "Crossing the Chasm" (Moore 1991, revised 1999 and 2014), which has sold over one million copies. "Crossing the Chasm" became a metaphor applied to enterprises with complex products to explain their struggle in the transition from early adopters to mainstream markets.

Moore used and expanded the diffusion of innovation (DOI) theory introduced by Rogers (2005) and revealed a chasm between the early adopters of a product (the technology enthusiasts and visionaries) and the early majority (the pragmatists). Moore showed that visionaries and pragmatists have very different expectations, which he characterizes as a "chasm" which needs to be crossed. His model and the suggested techniques to cross the chasm have had a significant and lasting impact on product launches until now (Schwabel 2013). "Crossing the Chasm" is closely related to the "technology adoption lifecycle" model (Beal and Bohlen 1957). The model consists of five main segments, which are recognized as (1) innovators, (2) early adopters, (3) early majority, (4) late majority and (5) laggards. Moore's theories applicable for disruptive innovations align with digital transformation approaches which assume emerging opportunities, but also rapidly changing conditions such as the increasing power of the consumers (Lemke and Brenner 2015, p. 197 f.). At the latest, when the replacement of an outdated ERP system is pending, the enterprises' decision makers are confronted with Moore's chasm: They need to decide between implementing a more modern version of a classical ERP system or coping with the new approaches related to digital transformation.

2.2 Systems of Engagement

Closely related to the "crossing the chasm" model is Moore's concept of Systems of Engagement (SoE), a class of communication-focused and collaboration-focused systems, which are an essential component of a new people-centric era (Moore 2011, 3). In prior decades, IT innovations usually emerged within large enterprises or public institutes and then made their way to medium and small enterprises until they finally reached the consumer markets. This sequence has turned around: Today, children and students lead innovations, followed by adults as well as small and medium-sized enterprises (e.g. Prensky 2001a, b; Roberts 2005; Koutropoulos 2011). Eventually, large enterprises adopt the new IT trends and innovations (Moore 2011, p. 2).

Moore postulates the "consumerization of IT" as a highly relevant factor in today's business. Two significant aspects appear. The first aspect relates to societal changes. The upcoming generation, the digital natives, are the future employees and customers. They are used to communicating and collaborating via mobile devices and to actively taking part in the Web 2.0. Related to their working life, they aim for self-realization and they expect easy-to-use tools, receiving information right at the time they need it (Lemke and Brenner 2015, p. 76).

The second aspect relates to potential productivity gains. The consumerization of IT can serve as the next significant driver for business growth and prosperity. In the past, powerful IS with comprehensive data repositories enabled enormous economic expansion (Ganesh et al. 2014, p. 6 f.). ERP systems, with inherent exhaustive data repositories, are one of the most prominent examples of these innovations. Therefore, and according to Moore (2011, p. 3), ERP systems can be referred to as "Systems of Records" (SoR) that paved the way for the automation of processes and outsourcing and led to operating efficiency and tremendous cost savings.

In the last decade, SoRs have more or less turned into a commodity, requiring enterprises to concentrate their resources on particular core competences in order to differentiate and sharpen their strengths. Further, products or services are no longer a result of one enterprise's activities as there is a network of partners that operates as one unit—a "boundless enterprise" (Picot et al. 2009 cited in Lemke and Brenner 2015, p. 217).

In the (imminent) digital age, collaboration between employees, divisions, suppliers and other internal and external parties is regarded as a key success factor to achieve the next wave of productivity gains. By utilizing SoE, employees will be enabled to deliver the required conversion in view of the new ways of collaboration. With their strong focus on social interactions, the younger generation supports the sharing of ideas and is used to working with and through social media. Therefore, future employees will have the potential to elevate collaboration to a higher level. Significant breakthroughs will take place by applying collaboration tools, which promise easy and user-friendly access and use. The facilitation of the new way of working requires the support of various technical formats such as texts, images, audio and video (Moore 2011, p. 5). Nevertheless, it is crucial to understand that, according

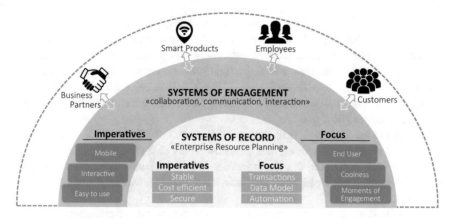

Fig. 1 Systems of engagement meets systems of records. (adopted from Corcoran 2013, p. 12; Matzke 2012; Lemke and Brenner 2015, p. 183)

to Moore, the "old" IS/ERP systems—the SoR—will not disappear. As such, SoE will enhance the existing enterprise IT landscape (Moore 2011, p. 6).

Figure 1 visualizes the relationship of both categories. The inner circle builds the SoR. This serves as the application backbone that automates essential business processes. SoR are required to run in a stable manner, to be cost effective and ensure high-level security and are built on a highly structured data model to support regular transactions. The SoE build a new layer, which is dedicated to interaction with users or smart services/products. SoE are designed to enable collaboration and focus on "Moments of Truth", the essential points during interactions. Hence, these systems have to be user-centric, focusing on mobile access and ease of use (Corcoran 2013, 12; Lemke and Brenner 2015, p. 183).

Enterprises that intend to incorporate SoE into their existing application landscape need to consider two things. First, more effort needs to be directed towards IT-enabled collaboration and people-centricity. Secondly, it is important to decide how SoE will be aligned with already existing SoR. The possibilities range from clear separation to full integration. We pick up both aspects in our model in Sect. 3.

2.3 Selective Applications

Moore (2011) discusses the concept of SoE using two perspectives. First, he makes recommendations for CIOs in B2B-oriented enterprises. The "Moments of Truth" appear in tasks such as holding meetings across different time zones as well as file sharing and recordings or solving complex business issues collaboratively. Suitable collaboration tools are mainly "Web 2.0 family" related, e.g. blogs or wikis, utilizing and combining the wisdom of experts and the crowd (Moore 2011, p. 7). Secondly, Moore turns to the CIOs working in B2C-oriented enterprises. In the past, consumers fulfilled their demands by actively searching for desired products/services in the World Wide Web. In the near future, B2C-enterprises will need to foresee the (potential) requirements of their customers and provide accurate and tailored information in the right place and at the right time. Powerful algorithms that enable predictive analytics help to achieve this. Further, it is recommended that SoE is integrated with the well-known SoR to support customer contact throughout several channels (Moore 2011, p. 8 f.).

While Moore deals with SoE in a generic context of B2B and B2C, the concept has been applied in other situations. There are many cases in the literature in which the SoE model is adapted in certain industries with different technical approaches (e.g. Dey et al. 2014; Book et al. 2017). In the following, we discuss some selected examples.

Example 1: In the banking sector it is extremely important to guarantee confidence, security and regulatory compliance (Corcoran 2013, p. 4). However, banks may be forced to respond to changed customer behavior and demands (Corcoran 2013, p. 9). The study suggests that banks should reshape their application architecture, including the separation of IT systems into SoR (which is aligned to internal business processes) and SoE (which takes care of customer interactions). The latter focuses on customer expectations and is therefore designed from an "outside-in" perspective (Corcoran 2013, p. 14 f.). As SoE need to be agile and easy to use, the study raises the point that integration with SoR has to be considered carefully. In cases where customer interaction does not depend on real-time exchange with SoR, it is valid to keep the systems separated. In this way, risks resulting from frequent SoR-interface changes are avoided (Corcoran 2013, p. 22).

Example 2: The adoption of SoE in the context of I4.0 is described by Wehle and Dietel (2015, p. 211 f.). They show the application of SoE focused on production processes and the optimization of maintenance activities. The challenges of integrating components like SoR, shop floor machines, sensors and mobile devices are investigated, and as a solution, a cloud-based architecture is designed. One core element is a network of connected sensors measuring variables such as air humidity or temperature. The sensors are part of the SoE, receiving real-time data about the physical environments and transfer it via cloud services to the existing SoR. The SoR is relating the sensor data to the information stored for a specific work unit (e.g. capacity, maintenance schedule). A correlation of sensor and SoR data helps to evaluate how the air quality affects the quality of the produced product.

Example 3: Li et al. (2014) use Moore's concept to investigate software-defined environments. The result is a framework that visualizes the potential evolution and consolidation of SoE and SoR (see Fig. 2). The model consists of two axes. The Y-axis shows that a certain class of software systems relies on agility and therefore needs continuous updates. The platforms used for these systems have to be scalable and adaptive. Agility is the main requirement on this axis. The X-axis points out that some software products are heavily dependent on an efficient technological platform. They have to run in a stable manner while guaranteeing high performance and strong compliance-related controls, such as ERP systems. Efficiency is the main factor on this axis.

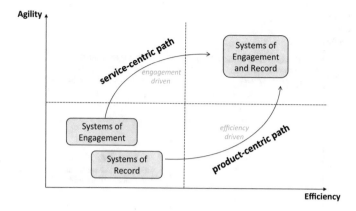

Fig. 2 Evolution of systems of engagement and systems of records. (adopted from Li et al. 2014, 1:6)

For example, SoE such as Facebook, Netflix and Twitter, follow the service-centric path when deciding on a technological platform. Agile cloud services provided by Amazon or Microsoft, for example, are usually the first choice. For SoR related categories, the product-centric path is more commonly applied. Platforms such as Oracle's Exadata or SAP's Hana ensure the required efficiency (Li et al. 2014, 1:6 f.).

The future vision anticipates a convergence of both perspectives, which is about to emerge, and in some cases, is already taking place. For example, many product-centric platforms (SoR; ERP systems) are working not "on premise" only, but also as agile cloud solutions e.g. SAP Hana. In the same way, cloud leaders such as Amazon are providing more and more computing power in order to increase efficiency (Li et al. 2014, 1:6 f.).

3 Model of ERP Transformation

In the following section, a model of ERP transformation based on the concept of SoE is developed. The final model leads to four categories showing the influence of SoE in view of future ERP initiatives.

Hevner et al.'s design science approach (Hevner et al. 2004; Hevner and Chatterjee 2010) demands that research needs to be relevant and solve a business issue. We showed that the majority of medium and large enterprises have already established SoR. In the operating phase, enterprises work on continuous improvement and/or on consolidation of SoR. However, there are still small and medium enterprises just starting to implement ERP systems (Leyh 2014, p. 1). How will the discussion around the SoE influence their future ERP initiatives? What options are possible? The designed "ERP transformation model" will be an artefact that targets these questions. It serves as a supporting tool for CIOs and ERP project managers during the pre-

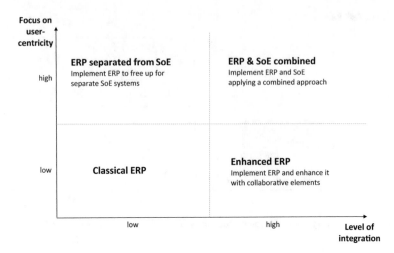

Fig. 3 ERP transformation model

project phase where general strategic directions for a new or next-generation ERP system are initiated.

Another important factor is Hevner et al.'s claim regarding rigor in science. To ensure rigor and transparency we use our expertise, the outcomes of the literature analysis, the application of a "theory", namely Moore's concept of SoE and the use cases (see Sect. 2). Based on all this, the newly created ERP transformation model separates the decision indicators for investing in ERP systems into four categories (see Fig. 3). The model is built on two dimensions:

Focus on user-centricity—a central paradigm of SoE is their focus on the "Moments of Truth"; the design of these systems is strongly aligned to the users as well as to smart products/services and their interactions. According to Moore, collaboration will be a key element to ensure success and productivity gains. Enterprises that implement or consolidate their ERP system/landscape could have a low or high level of interest in this topic.

Level of integration—enterprises that have already established SoE, need to leverage collaboration. As long as core business processes are executed within ERP systems, they should be aligned with existing SoE initiatives. Enterprises need to decide if they are willing to invest in an integrated ERP and SoE approach or if they tend to keep both worlds separate.

The two dimensions—people-centricity and integration—build the frame for the following four categories:

Classical ERP—Enterprises use an ERP system in order to achieve more "classical" objectives such as the automation of process steps, the harmonization of processes and/or IT systems or to support outsourcing. The main drivers are the reduction of costs, the shortening (throughput) of times or quality improvement. This quadrant stands for low levels of both integration and people-centricity.

ERP separated from SoE—If an enterprise has a clear focus on collaboration and aims to invest in SoE then it might use the ERP system that concentrates on process automation and relieves employees of routine tasks to allow more interactive tasks. In this constellation, SoE is a separate approach, not directly aligned with (new) ERP implementations or continuous improvement initiatives. Especially for enterprises in highly regulated industries (e.g. pharmaceuticals or banking), this might be a valid approach that permits the move towards a user-focus without impacting on existing SoR landscape (see Example 1 in Sect. 2.3). This quadrant stands for a growing level of people-centricity and a low level of integration.

Enhanced ERP—This applies when an enterprise aims to use ERP systems in a classical manner and tries to enhance them with collaborative elements related to SoE from an inside-out perspective. Collaborative elements are added without influencing core functionalities, and are aligned with the ERP-related processes. In addition, this could be a knowledge management tool, possibly related to mobility and/or other engaging elements. Further, the ERP vendors themselves have an increased interest in enriching their products with tools and functionality out of the SoE category in order to attract more customers (El Kadiri et al. 2016, p. 28). This quadrant stands for a growing level of integration and a low level of people-centricity.

ERP and SoE combined—Enterprises might be willing to "cross the chasm" and significantly change the way they build up/automate their (core) processes. For them, it is not sufficient just to enhance ERP systems or to add SoE approaches to the IT landscape. More than this, they aim to achieve a paradigm shift in the direction of digital transformation. Consequently, the individual people (who often miss collaboration opportunities)—customers, employees, business partners—are encouraged to take center stage. Applying a people/user-focused approach, an investigation is conducted regarding which IT tools and support are required at which point in time. ERP related processes and SoE are integrated in a joint ecosystem to fulfill these needs. This quadrant stands for a growing level of integration and people-centricity. One possible example for this quadrant could be a cloud-based ERP platform where enterprises could build their own ERP system by composing suitable Web Services, possibly offered by different software vendors. Chen et al. (2015) elaborated this idea, although they did not set their focus on user-centricity or design the ERP system along the relevant "Moments of Truth".

The ERP transformation model (see Fig. 3) shows four different strategies when deciding about the future ERP landscapes. From our point of view, the category "ERP and SoE combined" offers the highest potential for future productivity gains. However, it is a challenging endeavor as integration comes along with an increased level of complexity. For this reason, Corcoran (2013, p. 22) suggests that it is a valid decision not to integrate SoE and SoR in certain cases. In addition to increased productivity, the combination of ERP and SoE serves as a driver to "cross the chasm" within the enterprise organization.

In the past, the classical ERP systems focused on the support of the operational level within an organization. In addition, the executive level benefitted from aggregated data and increased transparency due to the harmonized data repository (Ganesh et al. 2014, p. 6 f.). The next-wave ERP combined with SoE will empower employees in the middle of the organization as is claimed by Moore (2011, p. 3). In addition to efficient operational processes and transparent information flows, an organization will be able to use and enlarge its knowledge base, and hence reach productivity gains, thanks to the support of agile communication, collaboration and interaction.

4   Conclusion

This last section outlines the relevance of the research interest, concludes the results and shows a derived research agenda for the next phase planned.

The first objective of the study focused on the relation between digital transformation and enterprises' ERP systems. After reviewing the literature, we identified a research gap and missing concepts regarding digital transformation aspects and "commodity" ERP systems. In addition, we found out that the generation of digital natives and their expectations are more comprehensive regarding collaboration and the use of social media—this needs to be considered more in the future.

The second objective aimed to discover a widely accepted, known and established theoretical model/concept as a foundation and decision template. Such a template seems necessary and valuable for enterprises that launch an initiative(s) regarding the new/next level of ERP (transformation). We decided to use Moore's concept of SoE (Moore 2011) as a starting point for discussions. We developed a new ERP transformation model (see Fig. 3), based on Moore's explanations (adopted in Fig. 1), the case studies applying the SoE concept (Sect. 2.3) and the model of Li et al. (2014) (see Fig. 2). The model is predestined to support decision makers, firstly to discover their current ERP-related situation and secondly, to define which of the four categories are appropriate to achieve the goals associated with their enterprise's digital transformation strategy. After setting the destination, the future course needs to be set in the direction of the chosen quadrant.

Regarding the present and future research agenda most of the analyzed literature focuses on digital transformation and the disruptiveness of new technologies. Only limited literature is available that specifically addresses ERP systems and their future as monolithic, more or less inflexible systems. This is remarkable as most large and medium-sized enterprises work with ERP systems.

The developed ERP transformation model (see Fig. 3) serves as an approach for a people-centric ERP transformation as well as a foundation for future research opportunities. It is the first milestone in a long-term research with the following agenda: we identified a need for research assigned to future-oriented and visionary models for ERP transformation. In addition, studies are needed that address the feasibility of these kinds of models in practice. Further research should aim to validate and use the developed ERP transformation model in practice. In addition, key success factors for ERP implementation/consolidation initiatives aligned to the digital strategies of enterprises need to be identified to determine relevant considerations. From the methodology perspective, more case study research, potentially using the case-based reasoning method, seems necessary to elaborate more examples regarding ERP transformation. Finally, the validation and application of the developed ERP transformation model to other areas of digital transformation may also be an interesting prospective field of further research.

References

Andelfinger VP, Hänisch T (eds) (2017) Industrie 4.0: wie cyber-physische Systeme die Arbeitswelt verändern. Springer, Wiesbaden

Bahssas DM, AlBar AM, Hoque MR (2015) Enterprise resource planning (ERP) systems: design, trends and deployment. Int Technol Manag Rev. 5(2):72–81

Baker MJ (2000) Writing a literature review. Mark Rev 1(2):219–247

Barkol O, Bergman R, Kasravi K, Golan S, Risov M (2012) Enterprise collective: connecting people via content. TechReport Hewlett-Packard Development Company. http://www.hpl.hp.com/techreports/2012/HPL-2012-102.pdf. Accessed 1 Jul 2017

Bauernhansl T, Hompel M, Vogelheuser B (eds) (2014) Industrie 4.0 in produktion, automatisierung und logistik: anwendung, technologien und migration. Springer Vieweg, Wiesbaden

Beal GM, Bohlen JM (1957) The Diffusion process: agricultural extension service. Iowa Agricultural and Home Economics Experiment Station Special Report

Benbasat I, Zmud RW (1999) Empirical research in information systems: the practice of relevance. MIS Quart 23(1):3–16

Book M, Gruhn V, Striemer R (2017) Gezähmte Agilität. In: Book M, Gruhn V, Striemer R (eds) Erfolgreiche agile projekte. Springer, Berlin, pp 3–15

Botthof A, Hartmann EA (eds) (2015) Zukunft der Arbeit in Industrie 4.0. Springer, Berlin

Chalons C, Dufft N (2016) The role of IT as an enabler of digital transformation. In: Abolhassan F (ed) The drivers of digital transformation. Springer, Cham, pp 13–22

Chen CS, Liang Y, Hsu HY (2015) A cloud computing platform for ERP applications. Appl Soft Comput J 27:127–136

Corcoran D (2013) Systems of engagement: apple usability, amazon agility, fort knox security. AIIA Whitepaper. https://aiia.com.au/documents/thought-leadership/financial-services-thought-leadership/2013/thought_leadership_papers_systems_of_engagement_apple_usability_amazon_agility_fort_knox_security_nov_2013.pdf. Accessed 01 Jul 2017

Dey P, Mohania M, Weldemariam K (2014) Some Issues in modeling user behavior data in systems of engagement. In: Ait AY, Bellatreche L, Papadopoulos GA (eds) Model and data engineering. MEDI 2014. Springer, Cham, pp 9–12

El Kadiri S, Grabot B, Thoben KD, Hribernik K, Emmanouilidis C, von Cieminski G, Kiritsis D (2016) Current trends on ICT technologies for enterprise information systems. Comput Ind 79:14–33

Finger J (2012) Erfolgreiche ERP-Projekte: Ein Rezeptbuch für Manager, 2nd edn. Springer, Berlin

Fosser E, Leister O, Moe C, Newman M (2008) ERP systems and competitive advantage: some initial results. In: 2nd 3gERP workshop. Frederiksberg, Denmark. http://3gerp.iwvi.uni-koblenz.de/docs/ERP_competitive_advantage_MSoftNov08.pdf. Accessed 01 Jul 2017

Ganesh K, Mohapatra S, Anbuudayasankar SP, Sivakumar P (2014) Enterprise resource planning: fundamentals of design and implementation. Springer, Cham Heidelberg

Hevner AR, March ST, Park J, Ram S (2004) Design science in information systems research. MIS Quart 28(1):75–105

Hevner AR, Chatterjee S (2010) Design science research in information systems. Design research in information systems. Springer, Boston, USA, pp 9–22

Kaufmann T (2015) Geschäftsmodelle in Industrie 4.0 und dem Internet der Dinge: der Weg vom Anspruch in die Wirklichkeit. Springer, Berlin

Köffer S, Ortbach K, Niehaves B (2014) Consumerization and Job performance: a theoretical framework for future research. Commun Assoc Inf Syst 35(14). http://aisel.aisnet.org/cais/vol35/iss1/14. Accessed 01 Jul 2017

Koutropoulos A (2011) Digital natives: ten years after. J Online Learn Teach 7(4):525–538

Kurbel KE (2013) Enterprise resource planning and supply chain management. Springer, Wiesbaden

Laudon KC, Laudon JP (2016) Management information systems: managing the digital firm. Pearson, London

Lemke C, Brenner W (2015) Einführung in die Wirtschaftsinformatik. Springer, Berlin

Leyh C (2014) Which factors influence ERP implementation projects in small and medium-sized enterprises? In: Twentieth Americas conference on information systems, Savannah, 2014

Li CS, Brech BL, Crowder S, Dias DM, Franke H, Hogstrom M, Rao J (2014) Software defined environments: an introduction. IBM J Res Dev 58(2/3):1:1–1:11

Mabert VA, Soni A, Venkataraanan MA (2001) Enterprise resource planning: common myths versus evolving reality. Bus Horiz 44(3):69–78

Matzke P (2012) Mobile Prozesse setzen neue Prioritäten für CIOs. https://www.computerwoche.de/a/mobile-prozesse-setzen-neue-prioritaeten-fuer-cios,2522019. Accessed 01 Jul 2017

Monk EF, Wagner BJ (2013) Concepts in enterprise resource planning, 4th edn. Course Technology Cengage Learning, UK

Moore G (1991) Crossing the chasm: marketing and selling technology products to mainstream customers. Harper Business, New York. ISBN 9780887305191

Moore G (1999) Crossing the chasm: marketing and selling technology products to mainstream customers, Rev. edn. HarperBusiness, New York. ISBN 9780066620022

Moore G (2011) Systems of engagement and the future of enterprise IT. A Sea Change in Enterprise IT. In: AIIM Whitepaper. http://info.aiim.org/systems-of-engagement-and-the-future-of-enterprise-it. Accessed 01 Jul 2017

Moore G (2014) Crossing the chasm, marketing and selling disruptive products to mainstream customers, 3rd edn. Harper Business, New York

Nievhaves B, Köffer S, Ortbach K (2013) The effect of private IT use on work performance—towards an IT consumerization theory. In: 11th International Conference on Wirtschaftsinformatik, Leipzig, Germany, 27 Feb–01 Mar 2013

Orosco C (2015) The future of enterprise IT: an interview with geoffrey moore (Part 1/2). Available via ForbesBrandVoice, 17th Mar 2015. https://www.forbes.com/sites/netapp/2015/03/17/geoffrey-moore-part-1/#797ecd287ef1. Accessed 01 Jul 2017

Panorama Consulting (2016) Everything you need to know about digital transformation. Available via Panorama Consulting Solution, Whitepaper, Denver. https://www.panorama-consulting.com/resource-center/erp-industry-reports/everything-you-need-to-know-about-digital-transformation. Accessed 01 Jul 2017

Picot A, Reichwald R, Wigand RT (2009) Die grenzenlose Unternehmung: Information, Organisation und Unternehmensführung im Informationszeitalter, 4th edn. Springer, Berlin

Prensky M (2001a) Digital natives, digital immigrants (Part I). On the horizon, vol 9(5). MCB University Press, pp 1–6

Prensky M (2001b) Digital natives, digital immigrants: do they really think different? (Part II). On the horizon, vol 9(6), MCB University Press, pp 1–6

Roberts G (2005) Technology and leaning expectations of the net generation. In: Oblinger DG, Oblinger JL (eds) Educating the net generation. Educause 3.1–3.7

Rogers EM (2005) Diffusion of innovations. Free Press, New York

Schwabel D (2013) Geoffrey moore: why crossing the chasm is still relevant. Available vai Forbes, 17 Dec 2013. http://www.forbes.com/sites/danschawbel/2013/12/17/geoffrey-moore-why-crossing-the-chasm-is-still-relevant/#26ec69dc782d. Accessed 01 Jul 2017

Schwenk M (2014) Weshalb ERP-Projekte im Mittelstand scheitern. HMD Praxis der Wirtschaftsinformatik 49(3):34–42

Seddon PB (2005) Are ERP systems a source of competitive advantage? Strateg Chang. 14(5):283–293

Sendler U (ed) (2013) Industrie 4.0: Beherrschung der industriellen Komplexität mit SysLM. Springer, Berlin

Sontow K, Schürmeyer M (2014) Industrielle Vernetzung verwandelt ERP Systeme. Is report (04), pp 18–23

Stefanou CJ (2014) Adoption of free/open source ERP software by SMEs. Information Systems for small and medium-sized enterprises. Springer, Berlin, pp 157–166

Tranfield D, Denyer D, Smart P (2003) Towards a methodology for developing evidence-informed management knowledge by means of systematic review. Br J Manag 14(3):207–222

Venkatesh V, Thong JYL, Xu X (2012) Consumer acceptance and use of information technology: extending the unified theory of acceptance and use of technology. Social Science Research Network, Rochester

Vom Brocke J, Simons A, Niehaves B, Riemer K, Plattfaut R, Cleven A (2009) Reconstructing the giant: on the importance of rigour in documenting the literature search process. In: Proceedings of the 17th European conference on information systems (ECIS), Verona, pp 2206–2217

Wehle HD, Dietel M (2015) Industrie 4.0 – Lösung zur Optimierung von Instandhaltungsprozessen. Informatik-Spektrum 38(3):211–216

Determining Information Relevance Based on Personalization Techniques to Meet Specific User Needs

Barbara Thönssen, Hans Friedrich Witschel and Oleg Rusinov

Abstract The support of workplace learning is becoming increasingly important as change in every form determines today's working world in the industry and public administrations alike. Adapting quickly to any kind of change is just one aspect. Another is dealing with the information relevant to this change. A recommender system for workplace learning was developed within the European funded project Learn PAd. Even if the information is filtered based on a learner's context with the help of the recommender, information overload remains a problem. It is not only the sheer amount of information but also the (often little) time for processing it that adds to the problem, time needed to assess the quality of the information according to its level of novelty, ambiguity, etc. Therefore, we enhanced the Learn PAd's recommender by implementing a personalization strategy to filter (recommended) information based on a learner's context. Our research work follows a design science research strategy and is evaluated in an iterative manner, first by comparing it to previously elicited user requirements and then through practical application in a test process conducted by the project application partner.

Keywords Workplace learning · Information overload · Personalization
Recommender system

B. Thönssen (✉) · H. F. Witschel
Institute for Information Systems, University of Applied Sciences and Arts Northwestern
Switzerland, Von Roll-Strasse 10, 4600 Olten, Switzerland
e-mail: barbara.thoenssen@fhnw.ch

H. F. Witschel
e-mail: hansfriedrich.witschel@fhnw.ch

O. Rusinov
Dukascopy Bank SA, Route de Pre-Bois 20, 1215 Geneva 15, Switzerland
e-mail: oleg.rusinov@dukascopy.com

1 Introduction

Workplace environments are becoming increasingly complex: Many knowledge workers have to digest increasingly larger amounts of information and have to deal with change—for example in the context of public administrations the changes of regulations—that occur at a fast rate. This situation requires constant learning and efficient ways of searching for and selecting relevant information.

Within the Learn PAd[1] EU research project, novel learning approaches and solutions for workplace learning have been introduced (De Angelis et al. 2015). Therefore, a recommender system was developed that considers a learner's context regarding tasks she or he has to perform in business processes, organizational knowledge about his or her position in the organization and, his or her current skill level and working experience (Emmenegger et al. 2016). Based on this, the Learn PAd recommender suggests learning help from experts who might be contacted for advice, from similar cases that might be taken into account or from task descriptions and learning experiences that are available in the Learn PAd Wiki and in conventional learning material like books, audio and video files. Thus, learners are proactively provided with the help that they (are likely to) require, without having to search for it.

However, over time recommending learning help based on context information is not sufficient because the amount of relevant information is increasing constantly. Thus, information overload becomes a problem in workplace learning even if the information has already been filtered, based on a learner's context. Moreover, it is not only the sheer amount of information but also the (often little) time for processing it that adds to the problem; time needed to assess the quality of information like the level of novelty, ambiguity, uncertainty, intensity, or complexity (Lincoln 2011). For instance, in the Learn PAd system, learners are encouraged to contribute reports on their own relevant experiences—because these reports may grow large over time, filtering the relevant sections would contribute to keeping the information load at bay.

Therefore, Learn PAd's context-based recommender has been enhanced by personalization in order to (further) restrict the amount of suggested learning help.

The applied research method is design science research (Hevner and Chatterjee 2010). Since context information (e.g. about business processes, user tasks he or she has to perform in business processes, organizational knowledge about his or her position in the organization and his or her working experience) is represented in an ontology the method is complemented by the approach of Grüninger and Fox (1995) for ontology design and evaluation.

The chapter is structured as follows: In Sect. 2, we introduce the main concepts of the Learn PAd project to provide the background of the study. In Sect. 3, we give an overview of related work. Then we detail our approach in Sect. 4. In Sect. 5, we provide details on the performed evaluation and conclude in Sect. 6.

---

[1] See http://www.learnpad.eu.

2 Background of the Study

In this section we briefly introduce relevant terms and concepts from Learn PAd, including an explanation about how we support learning during business processes, how processes are mapped to a Wiki representation and how knowledge workers are supported by and contribute to the Wiki to improve models and thus achieve organizational learning.

The Learn PAd project introduces a novel model-driven approach for workplace learning in the application domain of public administrations (PA) (Emmenegger et al. 2016). It creates a collaborative environment in the form of a Wiki to represent business processes and tasks, to recommend learning help (e.g. the contact details of experts, selection of similar and conventional learning material like books, audio and video files) and permits tips and advice to be contributed by writing experience reports (Pierantonio et al. 2015).

To model collaborative workplace learning centered on business processes and their context, we extended existing meta models, e.g. standard notations such as Business Process Model and Notation (BPMN) (OMG 2011) and Business Motivation Model (BMM) (OMG 2010) and created new ones, based on standards (e.g. the Competency Meta Model is deduced from the European Qualifications Framework (EQF) (European Commission n.d.)). We then transformed the models and relations between them into an ontological representation for automatic reasoning and into Wiki pages and links for learning purposes. If a user (learner) logs into the Learn PAd Wiki and accesses a page with a task description, based on context information (about the process the task belongs to and the user), appropriate learning help is determined and recommended to the learner. Thus, the Wiki is used to represent information to learners in a convenient way and the ontology is used to infer the necessary context information. As the Wiki is meant to encourage users to contribute their knowledge to improve business performance (Emmenegger et al. 2016) experience reports may become rather long, reflecting many different aspects. Over time, users will face the situation that although a recommended experience report is highly relevant, reading through all the provided information becomes too time-consuming. The same applies of course for other learning materials or, for example, regulations: Often only a small fraction of their content is really relevant to the task at hand. For this reason, the recommender was enhanced by another function to permit extraction or highlighting of specific sections of the report.

3 Related Work

Reduction of information overload in workplace learning can be achieved in several ways, some of which have already been well studied by other researchers.

For instance, much research has been done in the area of information retrieval, i.e. for scenarios in which a learner searches actively for missing information. In many adaptive information retrieval approaches, the increased relevance of search results is targeted via personalization techniques—the relevance of retrieved results may be (re-)assessed based on interests, tasks and previous knowledge of learners:

- *Individualization/personalization of search results* can be based on general user interests, which might be modeled in different ways (Gauch et al. 2007). Specifically, they might be derived either automatically, based on a user's behavior in previous search tasks (Morita and Shinoda 1994; Shen et al. 2005) or modeled explicitly as, for example, in Hopfgartner and Jose (2014). There are also hybrid approaches that try to use explicitly modeled information about user preferences where possible, but fall back on implicit information when explicit information is unavailable (Fernández et al. 2011).
- *Context modeling* takes into account the fact that user interests are not generic, but may vary according to the context and task at hand. Thus, some approaches subdivide user profiles according to certain (search) tasks, thereby relying on models of the domain of discourse (Vallet et al. 2007; Mylonas et al. 2008). In Asfari et al. (2009), context is modeled explicitly as the task/activity that a user is performing.

Another aspect of information overload is the challenge of assessing the (probable) relevance of presented search results. Instead of forcing a searcher to access and read document content, a search engine can assist in this challenge via *query-biased document summaries*, which give a fast overview of the context in which query terms appear in a document. Computation of such summaries can be formulated as a sentence selection task and may rely on learning-to-rank approaches (Wang et al. 2007), classification (Metzler and Kanungo 2008) or on graph-based scoring of sentences based on sentence similarity graphs (Varadarajan and Hristidis 2006).

Instead of requiring a user to engage in active search activities, the *recommendation* of suitable information items aims at a proactive delivery of information to a user when that information is needed. In general, recommenders are often used in e-commerce where they aim at suggesting items/products to a user that might be of interest to that person. Recommenders can be based on *collaborative filtering* (Linden et al. 2003), i.e. using the preferences of other users with similar interests. They can also be *content-based*, i.e. based on a comparison between user profile and content or description of the recommended items. Or they can be based on *association analyses* (Sarwar et al. 2000) that identify which items are often picked together and use that knowledge to suggest items to be added to an existing selection.

In the context of learning, recommenders have been proposed as part of *Adaptive Educational Systems (AES)* (Brusilovsky and Peylo 2003). Such systems aim at recommending to a learner which learning materials to consume next or which action to perform next (Zaíane 2002), based on the current level of knowledge, learning style/preferences (Peter et al. 2009) and previous learning activities of that learner. More precisely, Hauger and Köck (2007) distinguish between *presentation support* on the one hand—additional explanations, based on what a learner already knows—and *navigation support*—recommending where to go next.

Hence, as proposed by Alshammari et al. (2014), an AES needs to have a *Learner Model*, comprising information about the learner's knowledge and skills, current learning goal and preferred learning style, and a *Domain Model* that represents the real-world domain to be studied by students, used to annotate learning materials to match against the learner model. The learner model then permits *personalized* recommendations, i.e. make them relevant for the learner in terms of both suitable level of knowledge and preferred learning style (Dunn and Dunn 1978). In terms of skill level, good recommendations are in a student's *zone of proximal development* (Vygotsky 1978), i.e. they help the student to gain skills that are just above his or her current level, but can be reached with appropriate guidance. Personalization of learning has also been studied extensively, e.g. for recommendation of learning paths (Chen 2008; Hwang et al. 2010).

As in other domains (see above), recommendations can be based on content similarity (Ghauth and Abdullah 2010) or collaborative filtering (Khribi et al. 2009).

Finally, there is (relatively little) work focusing on the recommendation of learning materials in workplace learning that considers the (current) work context of a learner. Notably, Abecker et al. (2000, 1998) and Schmidt and Winterhalter (2004) suggest that recommendations in workplace learning should be based on both the role and position of the user within the organization and on the (current) work context in terms of the task or process to be executed. Such information can be gained e.g. via the integration of information delivery with a workflow engine (Abecker et al. 2000). Another example of task-specific information delivery is provided by Ye and Fischer (2002), suggesting using the task context in a software development context to deliver relevant code to be reused in a programming task. In the Learn PAd project the learners' context, for example, their working environment, skills, experiences and personal preferences, were considered for recommendations (Emmenegger et al. 2016).

Overall, previous research on text summaries has focused on the search task. Proactive information delivery in workplace learning using task context for the production of short document summaries in a proactive information delivery approach has not been studied.

We address this research gap by designing a new solution for proactive, personalized and context-aware information delivery with "task-biased" text summarization.

4 Determining Information Relevance

In the following sections, we introduce the conceptual approach and describe the implementation. The methodology our research follows is design science research (DSR) (Hevner et al. 2004). In order to elicit realistic requirements and to make sure that the solution is applicable in practice, we created an application scenario together with partners from the Learn PAd project that reflects a user's real working environment and illustrates role and output of the enhanced recommendation system.

4.1   Example Scenario

The example scenario we introduce is part of a larger scenario investigated within the Learn PAd project called "European Project Budget Reporting" (EPBR). EPBR refers to the activities that an Italian public research body (in reference to its administrative offices) has to carry out if it participates in a European research project. As the process is rather complicated, many recommendations are provided to support learners, e.g. novice users, in performing it. Based on answers gathered in semi-structured interviews during the awareness phase of DSR the Wiki page reporting learners' experience is considered the most vulnerable to the information flood, along with lengthy learning materials, such as (new) regulations.

We imagine a situation where a novice user who recently started work for an organization that is partner in an EU-project consortium is asked to do the periodic financial reporting. Although many helpful recommendations to support the administrative reporting activities are provided, the newcomer is overwhelmed by the information she or he finds in the Wiki pages and may also not be able to find the relevant sections in the provided learning materials, e.g. regulations. She or he would appreciate a (relevant) subset of the information according to her or his personal profile (e.g. with respect to skills, experiences, acquired knowledge in the domain, learning preferences) and the specific case at hand. To meet his or her needs, the Wiki pages should only display sections that match his or her profile and condense details to the most relevant facts.

4.2   Concept

In the context of our example scenario, we address two important forms of textual information items that are accessible from task-related Wiki pages:

- *Learning materials*: here, we refer to any items that contain information related to a task, and which may support a performer of that task in finding solutions and making appropriate decisions within the task. In other words, we understand the term "learning materials" in a very broad sense, and do not assume that the materials have a specific didactic purpose. They may take the form of e.g. explicit guidelines or background information such as law texts, regulations or interpretations of these. Figure 1 shows an example of a (passage-segmented) EU regulation that is used as learning material.
- *Experience pages*: Here, we refer to special Wiki pages that contain user-contributed experience in textual form, as shown in the example in Fig. 2.

Dealing with information overload in such a context means relieving the user of the task of scanning vast amounts of contents of learning materials and/or experience pages, because information which satisfies a specific user's need is hidden in text that is not relevant within the specific situation.

> **Can part of the management tasks be performed by other beneficiaries?**

> Coordination tasks are part of the "management tasks"; however, "management tasks" include tasks beyond
> tasks will be performed by other beneficiaries and they will be reimbursed at 100% provided they comply w
> on financial statements). In certain cases (i.e. big projects) there could be in a project a beneficiary carrying

> **Can there be a scientific coordinator other than the Coordinator?**

> The coordinator in the GA is defined only by the tasks mentioned in Article II.2.3. Tasks related to the coor(
> is possible that this beneficiary in charge of the task of scientific coordination, may be internally (i.e. within
> another beneficiary of the ECGA. It will not be considered as the project coordinator. The tasks of scientific
> "research and technological development activities" (i.e. 50% /75% reimbursement rate). By their nature (sc

> Example:

> Beneficiary "B" is leader of Work Package I in Project X, and in charge of the publication of a competitive (
> Packages of the project. He also has to provide a certificate on the financial statements.

> Reimbursement rates:

> - For its RTD work: 50% (75% if falling under the cases detailed in Article II.16.1.2 of ECGA)
> - For its management work related to the competitive call within Work Package I: 100%
> - For its scientific coordination of the project: 50/75% (as this is part of the RTD activities)
> - For its management costs related to the certificate on financial statements: 100%

> **Can a financially weak legal entity be coordinator of a project?**

Fig. 1 A (passage-segmented) EU regulation as a specific example of learning material

We therefore propose filtering and summarizing the textual contents in order to highlight only those passages of text that are relevant in the context of a given task and that match the user's profile in terms of previous knowledge required and learning style.

To this end, we will construct a complex query which describes the user's task context and profile to improve the approach for building recommendations implemented in the Learn PAd project (Silingas et al. 2015).

More precisely, the query will be composed of:

- Important keywords that are extracted from the title and description of the task at hand. This information is contained in the respective business process model and its corresponding Wiki page. The keywords will be used to bias text summaries towards the task context.
- Keywords extracted from the description of the case (i.e. process instance) the user is working on. This data is available from an integration with a workflow engine, which captures the case-specific data. In our example scenario, this might comprise, for example, the EU funding program or the country for which the

Maria franca Fissolo Experienced Beginner - 2016.03.02

Tasks which can be subcontracted and conditions

In an FP7 project, a beneficiary (**university**) subcontracts task X for an amount of EUR 50,000. If this amount is below the threshold set by its national public rules (i.e. EUR 100,000), then the subcontract must comply at least with the conditions set out in the GA, even if the national rules do not set out any specific requirement.

No rating Add review

David H. Koch Experienced Beginner - 2016.02.07

Can depreciation costs for equipment used for the project but bought before the start of the project be eligible?

Equipment bought in January 2005, with a depreciation period of 48 months according to the beneficiary accounting practices. If a GA is signed in January 2007 (when 24 months of depreciation have already passed), and the equipment is used for this ECGA, the beneficiary can declare the depreciation costs incurred under the project for the remaining 24 months in proportion of the allocation of the equipment to the research project.

No rating Add review

Mark Zuckerberg Novice - 2016.01.13

VAT

VAT could be considered as a cost by the accounting of a beneficiary, but this cannot be used to claim it as an eligible cost with an FP7 project, as VAT is not an eligible cost (article II.14.3.a)

No rating Add review

Fig. 2 Example of a user experience page

financial report is made. The extracted keywords will help to make text summaries relevant to the particular case the user is working on.

- The acquired skills and preferred learning style of the learner (e.g. whether she or he learns better through visual or auditory stimuli etc., (see Dunn and Dunn 1978)). This information can be used to filter and rank results such that the user primarily sees items that match the preferred learning style and that are appropriate in terms of his or her skill level—i.e. leading to a level of knowledge that is just above the user's current level (see Emmenegger et al. 2016).

Obviously, our resulting approach is hybrid, being based on keywords to describe the task context and the explicit ontology-based modeling for user profiles. The next section describes in more detail how the queries are constructed and processed.

4.3 Implementation

Next, we present our algorithm TBPTSL—task-biased and personalized text summarization for learning.

As a starting point, we build on the Learn PAd approach in which a business process such as the one from our example scenario was modeled and this model was then translated into a series of Wiki pages that contain titles, descriptions and other attributes of model elements (e.g. tasks) (De Angelis et al. 2015). Furthermore, we assume that learning materials were attached to some of the Wiki pages and that users contributed their experiences to some of the tasks of the process in textual form on the experience pages.

Figure 3 shows an overview of the algorithm which we will explain in detail here:

Step 1: Index preparation: Before proactive information delivery can start, it is prepared by segmenting learning materials and experience page contents into passages and creating an index for these.

After the index has been prepared, the system is ready to support a user working on a case (i.e. a process instance), in our case, preparing a report for the EU Commission. The user enters the case data and is then taken through the various tasks of the process on the Wiki pages. Let us assume now that the user is working on a particular task within the process and has navigated to the corresponding page in the Wiki.

Figure 4 shows such a situation: The user is working on the task "Calculate direct cost" and sees a short description of that task, including recommended subtasks, a link to the experiences that other users contributed, at the bottom of the page. On the

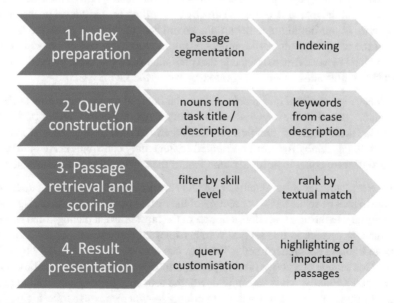

Fig. 3 Overview of the TBPTSL algorithm

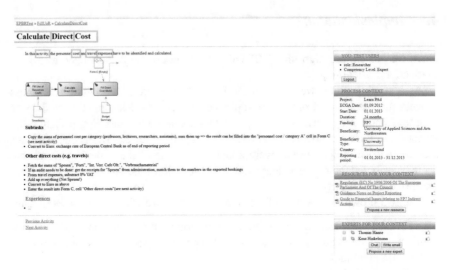

Fig. 4 Screenshot of a Wiki page representing a task and a case context in the Learn PAd Wiki

right hand side, the "Process Context" pane contains information about the current process instance, namely details of the project for which the financial report has to be created.

Step 2: *Query construction*: In this phase, the query will be constructed as described in the previous section, i.e. by extracting nouns from the title and description fields of the task elements, by capturing keywords from the description of the current case (e.g. "FP7" in the example of Fig. 4) and by forwarding the skill level of the user to the system. In our initial implementation, we did not consider the preferred learning style of the user; this will be the subject of future work. The yellow boxes in Fig. 4 show which keywords are extracted in that particular example. Note that information about skills, preferences, etc. is captured as part of the Learn PAd project.

Step 3: *Passage retrieval and scoring*: Next, the indexed passages are now retrieved and ranked. Based on the keywords of the query, a ranking of passages is performed using the BM25 retrieval function (Robertson and Walker 1994). Then the results are filtered: Especially for user-contributed content, only contributions of other users with a skill level equal to or higher than that of the current user will be retained.

Step 4: *Result presentation*: Finally, the results are presented as shown in Fig. 5 (note that due to lack of space the lower section of the result page is shown on the right side of the figure): In the upper section, snippets of the most relevant passages can be seen. A learner may check in this section whether relevant suggestions are available and navigate to these passages directly. The user can also customize the query by either adding (see orange box above the snippet) or removing keywords (orange box below the snippets). Finally, when scrolling down, the user sees the original content of the page—in this case user-contributed experiences—where passages are highlighted that were ranked highly in step 3. Thus, the user sees the passages in their original

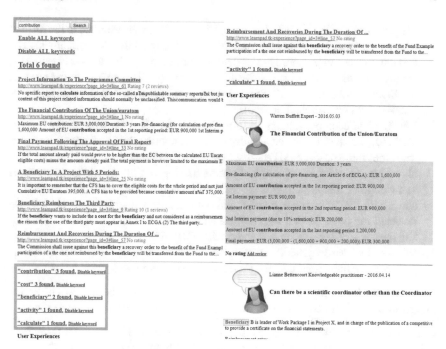

Fig. 5 Presentation of search results

context, but can scan the document much faster by focusing only on the parts that are relevant in the current task and case context.

5 Evaluation

5.1 Experimental Setup

In order to assess the usefulness of our proposed approach, we designed a small-scale experiment as follows:

1. We recruited four test persons who were vaguely familiar with the European Project Budget Reporting (EPBR) process.
2. We prepared a task to be performed by test persons within the context of the EPBR process. The task consisted of six subtasks (issues) that had to be performed as part of the tasks "Calculate direct cost" and "Calculate indirect cost"—issues typically consisted of deciding about the eligibility of certain costs or of computing costs, according to the EU guidelines. These tasks had to be performed for a (partially fictitious) research project case that we constructed.

3. We prepared a testbed in which extensive information was provided in the form of user-contributed experiences (which we extracted from the EU guidelines). The content covered many more topics than were needed to solve the six issues—i.e. the problem of information overload was present in this scenario. In the testbed, the functionality of our TBPTSL algorithm—as described in the previous sections—could be switched on or off.
4. We then had each participant perform three of the six tasks without the help of TBPTSL, i.e. just using the standard search functionalities of a web browser. The other three tasks could be solved using the full functionality of the TBPTSL algorithm. In order to reduce the "person bias" given by the different previous knowledge of the participants, we let the first two participants solve issues 1, 2 and 3 with TBPTSL and 4, 5 and 6 without it. For the other two participants, we reversed the order so that they started with issues 1, 2 and 3 without TBPTSL and then worked on issues 4, 5 and 6 with TBPTSL.
5. We asked all participants to try to solve the issues as quickly and accurately as possible; we then measured the time they needed to do so.

After completion of the tasks, we asked the participants about their subjective judgment of the TBPTSL functionalities using the following interview questions:

- In your opinion, does TBPTSL help you find the required information faster?
- In your opinion, does TBPTSL help you deal with information overload?
- Which changes would you suggest to improve TBPTSL?

5.2   Results

Table 1 shows the time needed by each test person for each issue. In the table, green cells represent an issue that was solved with the help of TBPTSL, whereas red cells stand for *not using TBPTSL*.

Table 1 Time needed by test persons to solve the 6 issues

Issue	TP1	TP2	TP3	TP4
1	4	20	2	5
2	3	15	4	3
3	3	10	6	5
4	5	15	4	1
5	4	15	3	1
6	6	15	4	7

We collected all answers to the interview questions and coded them manually. The results are as follows:

- Regarding the first question, there is no clear opinion among participants: Three of them agreed that it was hard to tell, two attributed this to the great variation in task difficulty ("it depends on the task") and one stated that the tasks were too easy to really notice a difference. Another participant also confirmed this, and claimed that for easy tasks the standard browser search functionality is sufficient. It should be noted though that in this context, an easy task is not one that is easily explained but one that can be searched for by using only one or two important keywords. In any case, these answers are quite consistent with the measured time for task completion.
- For the second question, there was total agreement among participants: All four stated that they believed that TBPTSL would help to reduce information overload. This indicates that participants see a potential for TBPTSL beyond the (rather easy) tasks that they performed.
- Lastly, each participant suggested a concrete improvement that was different from that the suggestion of the other participants. However, all suggestions had something to do with advanced information retrieval techniques: Participants proposed including phrases in the search (and hence index), to use stemming and to allow user ratings of contents and use them for ranking. We implemented all of these suggestions into our final solution. A last suggestion was to consider term semantics—which would require the use of a thesaurus or similar, but which we have not (yet) implemented.

It is obviously difficult to conclude much from the figures in Table 1 although the average time needed to solve issues with TBPTSL (all green cells, 6.25 min) is less than that of solving them without TBPTSL (all red cells, 7.08 min), the difference is not significant. This is of course due to the small sample size. In addition, there were other sources of variation that we could not fully control: The difficulty of the tasks is of course not completely comparable, neither is the prior knowledge of the participants. Therefore, more investigation will be needed to confirm the positive effect of TBPTSL in quantitative terms.

6 Conclusions and Future Work

Experiences reported in Wiki pages are a helpful instrument for learners who seek support for the execution of a (complex) business activity. However, when the Wiki is used a great deal, information/experience reports may become rather long and confusing and pages with explanations may become very detailed. What is commonly known as information flood can also be considered in this context. Therefore, it is not enough to recommend appropriate (Wiki) pages; the text also has to be extracted and summarized according to the user's/learner's profile. With our approach, we were able to show how recommendations for workplace learning can be improved by applying appropriate filters to text passages (in Wiki pages) based on task context and personal information. As we already represent background information in an ontology (a user's working context, e.g. the business activity she or he is about to execute, the organization structure of his or her organization) and his or her personal profile, we can enhance it by adding information about recommendations "liked" by the user If, for example, a learner "likes" contributions of a certain author more than of others, learning material related to this author might be recommended first. Hence, a promising thread in future work could be to analyze users'/learners' opinions of and experiences with learning material and use them to further improve recommendations, respectively filter recommended information.

References

Abecker A, Bernardi A, Hinkelmann K, Kühn O, Sintek M (1998) Toward a technology for organizational memories. IEEE Intell Syst 13:40–48. https://doi.org/10.1109/5254.683209

Abecker A, Bernardi A, Sintek M, Hinkelmann K, Ku O (2000) Context-aware, proactive delivery of task-specific information: the KnowMore project. Inf Syst Front 2:253–276. https://doi.org/10.1023/a:1026564510897

Alshammari M, Anane R, Hendley RJ (2014) Adaptivity in e-learning systems. In: 2014 Eighth International Conference on Complex, intelligent and software intensive systems (CISIS), pp 79–86

Asfari O, Doan B, Bourda Y, Sansonnet J-P (2009) Personalized access to information by query reformulation based on the state of the current task and user profile. In: Third International Conference on advances in semantic processing, 2009. SEMAPRO'09, pp 113–116

Brusilovsky P, Peylo C (2003) Adaptive and intelligent web-based educational systems. Int J Artif Intell Educ 13:159–172

Chen C-M (2008) Intelligent web-based learning system with personalized learning path guidance. Comput Educ 51:787–814

De Angelis G, Pierantonio A, Polini A, Re B, Thönssen B, Woitsch R (2015) Modelling for learning in public administrations—the learn PAd approach. In: Domain-specific conceptual modelling: concepts, methods, and tools. Springer

Dunn R, Dunn K (1978) Teaching students through their individual learning styles: a practical approach. Reston Publishing Company

Emmenegger S, Hinkelmann K, Laurenzi E, Thönssen B, Witschel HF, Zhang C (2016) Workplace learning—providing recommendations of experts and learning resources in a context-sensitive and personalized manner. In: Proceedings of special session on learning modeling in complex organizations (LCMO) at MODELSWARD'16

European Commission European Qualification Framework (EQF)

Fernández M, Cantador I, López V, Vallet D, Castells P, Motta E (2011) Semantically enhanced information retrieval: An ontology-based approach. Web Semant Sci Serv agents world wide web 9:434–452

Gauch S, Speretta M, Chandramouli A, Micarelli A (2007) User profiles for personalized information access. In: The adaptive web. Springer, pp 54–89

Ghauth KI, Abdullah NA (2010) The effect of incorporating good learners' ratings in e-Learning content-based recommender system. Educ Technol Soc 14:248–257

Grüninger M, Fox MS (1995) Methodology for the design and evaluation of ontologies. Ind Eng 95:1–10

Hauger D, Köck M (2007) State of the art of adaptivity in E-Learning platforms. In: Brunkhorst I, Krause D, Sitou W (eds) 15th workshop on adaptivity and user modeling in interactive systems

Hevner AR, March ST, Park J, Ram S (2004) Design science in information systems research. MIS Q 28:75–105

Hevner A, Chatterjee S (2010) Design research in information system - theory and practice. In: Voß S, Sharda R (eds) Integrated series in information systems

Hopfgartner F, Jose JM (2014) An experimental evaluation of ontology-based user profiles. Multimed Tools Appl 73:1029–1051

Hwang G-J, Kuo F-R, Yin P-Y, Chuang K-H (2010) A heuristic algorithm for planning personalized learning paths for context-aware ubiquitous learning. Comput Educ 54:404–415

Khribi MK, Jemni M, Nasraoui O (2009) Automatic recommendations for E-Learning personalization based on web usage mining techniques and information retrieval. Educ Technol Soc 12:30–42

Lincoln A (2011) FYI: TMI: Toward a holistic social theory of information overload. First Monday 16:1–15

Linden G, Smith B, York J (2003) Amazon.com recommendations: item-to-item collaborative filtering. IEEE Internet Comput 7:76–80. https://doi.org/10.1109/mic.2003.1167344

Metzler D, Kanungo T (2008) Machine learned sentence selection strategies for query-biased summarization. In: SIGIR learning to rank workshop, pp 40–47

Morita M, Shinoda Y (1994) Information filtering based on user behavior analysis and best match text retrieval. In: Proceedings of the 17th annual international ACM SIGIR conference on Research and development in information retrieval, pp 272–281

Mylonas P, Vallet D, Castells P, Fernández M, Avrithis Y (2008) Personalized information retrieval based on context and ontological knowledge. Knowl Eng Rev 23:73–100

OMG (2011) Business process model and notation (BPMN V 2.0)

OMG (2010) Business motivation model

Peter SA, Bacon E, Dastbaz M (2009) Learning styles, personalisation and adaptable e-Learning

Pierantonio A, Rosa G, Silingas D, Thönssen B, Woitsch R (2015) Architectures for business processess in organizations. In: Proceedings of the project showcase (PS'15), software technologies: applications and foundations (STAF'15). CEUR

Robertson SE, Walker S (1994) Some simple effective approximations to the 2-Poisson model for probabilistic weighted retrieval. In: Proceedings of SIGIR '94, pp 232–241

Sarwar B, Karypis G, Konstan J, Riedl J (2000) Analysis of Recommendation Algorithms for e-Commerce. In: Proceedings of the 2nd ACM Conference on Electronic Commerce. pp 158–167

Schmidt A, Winterhalter C (2004) User context aware delivery of e-learning material: Approach and architecture. J Univers Comput Sci 10:28–36

Shen X, Tan B, Zhai C (2005) Implicit user modeling for personalized search. In: Proceedings of CIKM '05, pp 824–831

Silingas D, Thönssen B, Pierantonio A, Efendioglu N, Woitsch R (2015) Business architecture for process-oriented learning in public administration. In: Business and dynamic change: the arrival of business architecture

Vallet D, Castells P, Fernández M, Mylonas P, Avrithis Y (2007) Personalized content retrieval in context using ontological knowledge. IEEE Trans Circuits Syst Video Technol 17:336–346

Varadarajan R, Hristidis V (2006) A system for query-specific document summarization. In: Proceedings of the 15th ACM international conference on Information and knowledge management, pp 622–631

Vygotsky LS (1978) Mind in society: development of higher psychological processes. Harvard University Press, Cambridge, MA

Wang C, Jing F, Zhang L, Zhang H-J (2007) Learning query-biased web page summarization. In: Proceedings of the sixteenth ACM conference on information and knowledge management. pp 555–562

Ye Y, Fischer G (2002) Supporting reuse by delivering task-relevant and personalized information. In: Proceedings of the 24th international conference on software engineering. pp 513–523

Zaïane OR (2002) Building a recommender agent for e-Learning systems. In: Proceedings of the international conference on computers in education. ieee computer society, p 55

Case-Based Reasoning for Process Experience

Andreas Martin and Knut Hinkelmann

Abstract The following chapter describes an integrated case-based reasoning (CBR) approach to process learning and experience management. This integrated CBR approach reflects domain knowledge and contextual information based on an enterprise ontology. The approach consists of a case repository, which contains experience items described using a specific case model. The case model reflects, on the one hand, the process logic, i.e. the flow of work, and on the other the business logic, which is the knowledge that can be used to achieve a result.

Keywords Case-based reasoning · Knowledge work · Business process management · Workflow systems · Enterprise ontology · Enterprise architecture Ontology-based case-based reasoning · Experience management Knowledge-intensive processes

1 Introduction

Knowledge and knowledge work are the key factors of successful companies nowadays. It is observable that there is a clear shift from routine work to knowledge work (El-Farr 2009). The term knowledge work "…indicate[s] the knowledge intensiveness of the current working tasks and the required abilities, skills, qualifications and

This chapter is the academic and peer-reviewed publication of results and some literal passages taken from the first author's Ph.D. thesis (Martin 2016).

A. Martin (✉) · K. Hinkelmann
Institute for Information Systems, University of Applied Sciences and Arts Northwestern Switzerland (FHNW), 4600 Olten, Switzerland
e-mail: andreas.martin@fhnw.ch

A. Martin
School of Computing, University of South Africa, Roodepoort, Johannesburg, South Africa

K. Hinkelmann
Department of Informatics, University of Pretoria, Pretoria, South Africa

working conditions for employees to accomplish their work" (El-Farr 2009, p. 3). Individuals who perform knowledge work act in different roles and ignite innovation in companies.

Conventional business process management (BPM) has been exceedingly successful for routine work. However, BPM and workflow management systems have deficiencies in dealing with the flexibility that is required to perform knowledge work. Classical workflow management systems are well suited to supporting the execution of rigidly structured business processes but fail to allow changes in unexpected situations (Adams et al. 2003).

Knowledge workers require the flexibility to determine their processes continuously—thus "making the knowledge worker's work more productive and focused, in addition to minimizing their stress and increasing interaction" (El-Farr 2009, p. 8). Knowledge work cannot be represented sufficiently in traditional business process management, where the work is structured and described in advance. It is especially difficult to predict future tasks because knowledge work deals with many different requirements simultaneously. Type and scope of tasks are hard to determine in advance, and the sequence of tasks may vary due to results already achieved and unforeseeable events. However, knowledge workers can benefit from business process management—even in knowledge work, it is possible to identify structured elements or process fragments. Gadatsch (2012, p. 43) distinguishes between three categories of processes: structured processes, cases, and ad hoc processes. Cases contain structured parts, but offer some degree of freedom for the workers. Through the combination of structured and unstructured elements within process orientation, knowledge work can be made more productive (Davenport 2005). Davenport (2010) looks at knowledge work processes in term of knowledge activities for the creation, distribution and application of knowledge. *This chapter describes a combined approach for maintaining process experience with the inclusion of case-based reasoning.*

Case-based reasoning (CBR) uses the knowledge of previously experienced cases to propose a solution to a problem. This research shows that CBR is adequate for retrieving, reusing, revising, retaining and storing functional knowledge and process knowledge of process instances or instances of process fragments. Consequently, the process experience must be made explicit and represented in a way that it is adequate for machines as well as for humans. The implication, then, is that the case description language should be as natural as possible to gain wide acceptance from the end users. The case description should also able to be processed by a computer using similarity and adaptability mechanisms during process execution. *This chapter presents a CBR approach for process experience, taking into account the requirements of both the users and the process execution systems.*

The structure of this chapter is as follows: It starts with Sect. 2, which describes the method and the research process, followed by Sect. 3, which introduces the application scenario. Section 4 shows our research contribution concerning the case-based reasoning approach including related work on CBR and ontology-based CBR (OBCBR). Further, Sects. 5 and 6 describe the two core elements of the approach, the case content and characterization models. Section 7 depicts a procedure model for this approach, and Sect. 8 provides an evaluation of the implemented artefact based on a specific application case. The article closes with the discussion, conclusion and future work, in Sect. 9.

2 Method

This research follows a design science research (DSR) strategy proposed by Vaishnavi and Kuechler (2004) as shown in Fig. 1. DSR *"…must produce a viable artifact in the form of a construct, a model, a method, or an instantiation…to important and relevant business problems"* (Hevner et al. 2004, p. 83). This research produces an *approach* (model), *instantiation* (prototype component), an ontology-based *case model* and a *procedure model* based on an application scenario (business case).

Besides, this research work is qualitative research, which demands a profound examination using evaluation data. Therefore, the evaluation data has been triangulated to ensure validity and provides a deeper understanding of a certain phenomenon (Cohen and Crabtree 2006).

Based on the research work of Cohen and Crabtree (2006), and with the help of Patton (2015, p. 661), it is possible to identify four major types of data triangulation:

1. Method triangulation with multiple data collection methods.
2. Source triangulation where different data sources from different time horizons or data capturing settings can be used.
3. Analyst triangulation with various observers and analysts.
4. Theory triangulation using different theoretical perspectives or theories.

The research was conducted using the admission process for master's students at the FHNW School of Business as the application scenario. For triangulation, first-hand data from the admission process was complemented with primary data based on interviews, documents and the study of artefacts. The evaluation of this approach was also ensured by evaluation data triangulation. In both cases, triangulation was ensured by various methods (method triangulation: interview, document and artefact studies and literature) and data sources (source triangulation).

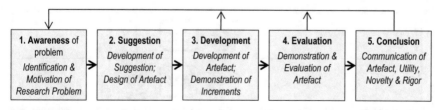

Fig. 1 Design science research process (adapted from Vaishnavi and Kuechler (2004) and Peffers et al. (2007))

3 Application Scenario

As mentioned before, the application scenario used here is the admission process for master's students at the FHNW School of Business. It was analyzed in detail by Martin (2016). The admission process is knowledge-intensive because the applicants can come from different universities and countries and have various degrees. Therefore, this scenario serves as an excellent application scenario in this research work. Figure 2 shows a fragment of the admission process in the Business Process Modelling Notation (BPMN) containing knowledge-intensive activities. In this admission process, a candidate has to be assessed as to whether he or she fulfils certain criteria in order to be qualified for the study course. These criteria involve a bachelor's degree qualification in a related field of study from an accredited institution, good or excellent grades, working experience, preferably in a related field of study and finally, good linguistic abilities in English.

The admission process begins when the online application of a prospective student arrives. This application is analyzed by a study assistant in the first activity "prepare eligibility check". The study assistant collects all the information needed to allow the dean of the program to check the eligibility of the candidate. If an applicant is assessed as eligible, the study assistant invites the applicant to an oral interview during the "invite for interview" activity. Otherwise, a rejection letter is sent. In this interview, the main goal is to validate the eligibility of the applicant. This activity is also highly knowledge-intensive. If, after the interview, the admission commission accepts the candidate during the "decide for acceptance" activity, an acceptance letter is sent to the candidate as an outcome of the "send acceptance letter" activity. If the candidate is not eligible, a rejection letter is sent.

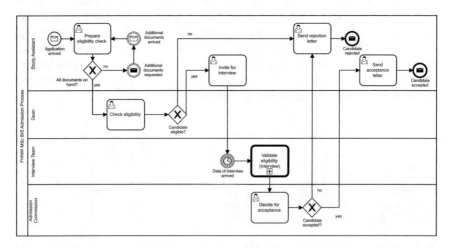

Fig. 2 Admission process for master's students at the FHNW school of business (adapted from Martin (2016))

Although the process seems to be very structured, it consists of knowledge-intensive tasks. For instance, the aim of the "prepare eligibility check" task is to determine whether an international bachelor's degree of an applicant qualifies for the master's program. Since the applicant may hold an international bachelor's degree, the study assistant can be confronted with a wide variety of degrees and certificates, which can be regarded as unknown if they are being verified for the first time. If a degree is unknown, the transcript of records is analyzed, and the accreditation of the awarding university is checked. As demonstrated in the application analysis by Martin (2016), the study assistant is confronted with a huge variety of cases. It is possible to learn from previous cases, which can later be adapted to new situations. In fact, the process participants already use a partly paper-based case base containing application knowledge. The process is knowledge-intensive due to a lack of decision logic, e.g. how to assess the equivalence of study programs, and a lack of procedural knowledge, i.e. which tasks to perform to obtain all decision-relevant data. Thus, the scenario demands a case base containing experiential knowledge, including process fragments explaining how a similar situation was resolved in the past.

4   Approach

Retrieving and maintaining existing knowledge and experience is an important aspect of different situations. This is especially the case when knowledge-intensive and agile activities occur. Case-based reasoning (CBR) can be an adequate method for retrieving experiential knowledge in an experience management system (Bergmann 2002).

4.1   Related Work—CBR and Ontology-Based CBR

As mentioned in the introduction Sect. 1, CBR can be seen as "reasoning by remembering" (Leake 1996, p. 2) or "reasoning from reminding" (Madhusudan et al. 2004) and as a technically independent methodology (Watson 1999) for humans and information systems. "*Case-based reasoning is both…the ways people use cases to solve problems and the ways we can make machines use them*" (Kolodner 1993, p. 27). Bergmann et al. (2009) regard CBR as a sub-field of artificial intelligence with its roots in cognitive science, machine learning and knowledge-based systems.

Figure 3 shows the main elements of a CBR approach. Traditionally, a case in CBR consists of a problem description and a solution description. Based on Bergmann (2002), this research work uses the term "characterization" for the problem description. "The case characterization part describes all facts about the experience that are relevant for deciding whether the experience can be reused in a certain situation" (Bergmann 2002, p. 48). Since the solution of a case can contain further elements such as a description of the experience, documents and process fragments, this research

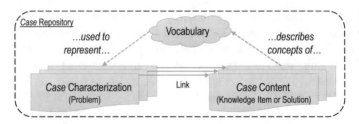

Fig. 3 CBR knowledge elements (adapted from Bergmann and Schaaf (2003))

work uses the term "case content". As additionally shown in Fig. 3, a CBR approach uses particular vocabulary to represent the case characterization and to describe the concepts of the case content.

To execute a CBR approach, knowledge to describe the experiences (the vocabulary) has to be acquired in advance. Therefore, it is advisable to provide the CBR with domain knowledge beforehand to reduce a knowledge acquisition bottleneck. *"The more knowledge is embedded into the system, the more effective [it] is expected to be"* (Recio-García et al. 2008, p. 54). This domain knowledge embedding is where ontologies can come into play. Ontologies can be used to implement the required CBR vocabulary. Combining ontologies with CBR enables the use of the power of ontologies in a CBR system. As discussed in Bergmann and Schaaf (2003) and Bichindaritz (2004), ontology-based systems can benefit from structural CBR and vice versa. Ontology-based CBR *"…can take advantage of this domain knowledge and obtain more accurate results"* (Recio-García et al. 2008, p. 54). Several frameworks exist for CBR inter alia COLIBRI (Bello-Tomás et al. 2004) and jCOLIBRI2 (Recio-García et al. 2014), myCBR (Bach and Althoff 2012; Thomas Roth-Berghofer and Bahls 2008) or CAKE (Bergmann et al. 2006; Maximini and Maximini 2007), which (partly) use ontologies to provide the CBR vocabulary.

4.2   Interlinked Case-Based Reasoning

Based on the analysis of the described application scenario in the previous section by Martin (2016), a case-based reasoning approach called ICEBERG was developed. This approach was instantiated into a demonstrator, which was applied at a software company (see Martin et al. 2013; Witschel et al. 2015 and Martin et al. 2017). The acronym ICEBERG stands for interlinked case-based reasoning, which is based on the commonly used iceberg metaphor in applied psychology, pedagogic and interpersonal communication. The ICEBERG approach was introduced together with an extension for process execution called ICEBERG-PE, which consists of the following four elements as depicted in Fig. 4 (Martin 2016, p. 115):

- The *case repository* is a central feature of the ICEBERG-PE approach, and it contains retained and learned cases.

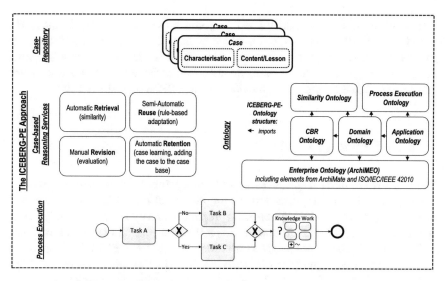

Fig. 4 ICEBERG-PE approach (adapted from Martin (2016))

- The *case-based reasoning services* provide the automatic retrieval of previous cases as well as semi-automatic reuse and adaptation of the previous cases to the current situation using rules, and the manual revision of new cases by the users. It also provides the automatic retention of cases with regard to adding the case to the case repository.
- The *ontology* is used for CBR configuration, storage of the enterprise and domain ontology and is simultaneously the vocabulary of the CBR approach.
- The *process execution* element is the instantiation of a business process based workflow engine running a workflow definition.

Figure 4 shows the heart of the ICEBERG-PE approach, the ontology structure. The ontology consists of several included ontologies that are formalized in RDFSPlus (Allemang and Hendler 2008), expressed in RDF(S) (W3C 2014a, b) and extended with elements of the OWL namespace (W3C 2012). The whole ontology is serialized in RDF 1.1 Turtle (W3C 2014c). When applying the ICEBERG-PE approach, the following five ontology structures can be identified (Martin 2016, p. 123):

- The CBR ontology consists of the configuration of the retrieval of the CBR system.
- The similarity ontology consists of concepts for creating a similarity model used by the retrieval mechanism.
- The domain ontology consists of enterprise-specific domain knowledge of an application scenario.
- The process execution ontology consists of concepts for interacting with a process execution engine and storing workflow relevant data.
- The application ontology specializes the domain ontology with respect to an enterprise idiosyncrasy as described by van Heijst et al. (1997).

In addition to this, the ICEBERG-PE approach uses the case viewpoint model introduced by Martin et al. (2017), which enables the users of the CBR system to have their own perspectives or views of the CBR model. This viewpoint model enables the users to define specific similarity values for a specific interest, which enables it in a figurative sense to provide several CBR systems in one system.

5 Case Content Model

Martin (2016) elaborated objectives for a case model for a CBR-based process execution approach. In summary, a case model for procedural knowledge should consist of *process fragments*, provide a *graphical representation*, have an *update functionality*, should contain *information resources* and provide the possibility to *model variants*.

The concept process fragments were introduced into the ICEBERG-PE approach during a requirements analysis with stakeholders. The stakeholders referred to this concept implicitly and explained it as a loose and unspecific fragmental element containing specifically performed activities, sub-process or case data. Eberle et al. (2009, p. 399) used this concept in a process modelling scenario as follows: "*Process fragments are reflecting the partial and intermittent knowledge one modeler [or a knowledge worker] has at a certain time about a specific situation.*"

Based on the considerations mentioned, the modeling language Business Process Feature Model notation (BPFM) (Cognini et al. 2015a, b) was selected as proposed by Cognini et al. (2016). Figure 5 shows the ICEBERG-PE system dialogue on the left side, including an example of a case content description that is modeled in BPFM. In addition to the BPFM-based case content, the following further elements are used to describe case interrelationships (Martin 2016, p. 165):

- Parent: The parent case element is used to express the sub-case relationship.
- Child: The child case element is the inverse of the parent element.

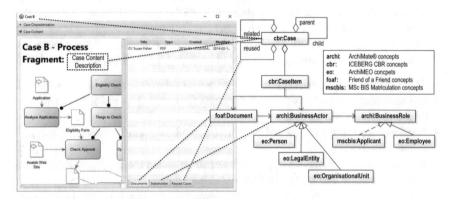

Fig. 5 Case content model elements (adapted from Martin (2016))

- Related: This element is used to express related cases.
- Reused: The reused case element is used to list tasks, which were reused in the adaptation phase of the CBR cycle.

In addition to this, Fig. 5 shows the inclusion of further information elements or knowledge items, which are documents and information about particular stakeholders who were involved in a case as described in Sect. 4.2.

6 Case Characterization Model

As mentioned in Sect. 4.2, the ICEBERG-PE approach relies on ontology-based CBR and uses an underlying ontology. The case characterization is used to describe (characterize) a case and to assess the similarity between cases using a retrieval mechanism and method. This case characterization is defined by the configuration of the similarity model (retrieval mechanism) including the case viewpoint model. Figure 6 shows an exemplary similarity model configuration including a viewpoint.

With the inclusion of an ontology structure in a CBR system, it is possible to "...*take advantage of this domain knowledge and obtain more accurate results*" (Recio-García et al. 2008, p. 54). The case characterization model has been extended with the following task specific elements based on Martin and Brun (2010) and Brander et al. (2011), to reflect the process execution context:

- The *task objective* element describes the goal of the task itself. This element has some similarities to the name and/or description of a BPMN activity.
- The *task role* element is used to describe the role of the person involved in the task. Through the inclusion of an enterprise or domain ontology, it is possible to reuse an existing enterprise-specific role/organizational model.
- The *task user* elements are used to indicate the person who described the case.

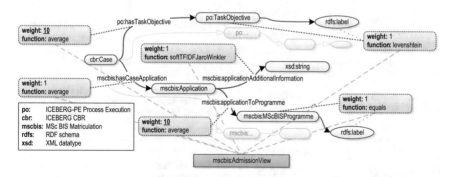

Fig. 6 Exemplary configuration of a case characterization including view (simplified and adapted from Martin (2016))

7   Procedure Model

As mentioned, the ICEBERG approach was elaborated over a longer period during which the results were described in several publications (Cognini et al. 2016; Martin et al. 2013, 2017; Witschel et al. 2015). This made it possible to derive a procedure model from practice for the implementation of the ICEBERG approach in a process execution context.

Such a procedure model serves as a basis for a future instantiation of the ICEBERG approach and, as Ju et al. (2016) emphasized, it provides a clear picture of the involvement of end-users and experts. The ICEBERG Procedure Model for Process Execution (Martin 2016, pp. 133–136), as shown in Fig. 7, consists of the following twelve steps:

1. To understand a particular current use case, a first *situation analysis* is performed by applying requirement engineering and creativity techniques.
2. Next, the *stakeholders*, their *concerns* and finally the *thing(s) of interest* are derived, which can be used to elaborate the mental models of the different stakeholders to ultimately derive the viewpoint-based similarity model as introduced by Martin et al. (2017).
3. Then, a generalized and overall *process model* is elaborated to provide an overview of a later process execution configuration.
4. It is advisable to create exemplary *case content* with initial data and discuss it with the stakeholders, and to define how further initial data is acquired.
5. Next, an initial analysis concerning process fragments needs to be conducted, since this approach reflects process execution and procedural knowledge. Consequently, a case modeling language (graphic or textual) or a subset of modeling language should be selected or tailored according to the previous complexity analysis, which is used as a *process fragment model*.

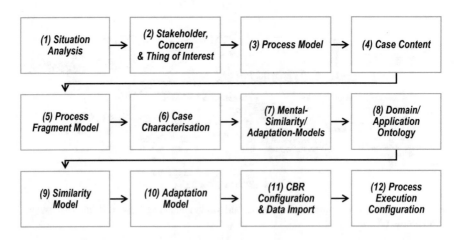

Fig. 7   ICEBERG procedure model for process execution (adapted from Martin (2016))

6. At this stage, it is essential that enterprise-specific elements such as enterprise models, enterprise-specific conceptualizations or nomenclature, or even an enterprise architecture can be used to build initial *case characterization* vocabulary.
7. The ultimate goal of any CBR implementation it to represent as adequately as possible *mental similarity* and *adaptation models* in a consolidated form as a configurable model. Before such a configurable model can be implemented, the various mental models need to be elicited from and consolidated by the stakeholders. The mental models are consolidated but are still dependent on the stakeholders and their concerns.
8. Based on the case characterization (step 6) and after defining the mental similarity and adaptation models, a specific *domain/application ontology* can be created for a process execution instantiation. This ontology creation is suggested with stakeholder involvement and is based on an enterprise ontology including application and domain concepts.
9. Based on the stakeholder-dependent mental models, the case characterizations, and the domain/application ontology, a CBR expert configures the *similarity model* dependent on a viewpoint-based similarity model as introduced by Martin et al. (2017). This configuration is made by determining global and local similarity functions and assigning weights.
10. Apart from the similarity model, the *adaptation model needs* to be configured by a CBR expert. This configuration can be made by defining manual or semi-automatic adaptation rules.
11. Subsequently, a CBR expert, who also ensures the initial *data import*, defines the *CBR configuration*.
12. Finally, a process engineer creates an executable and generalized process model and interlinks the CBR system with the process execution system in a *process execution configuration* step.

8 Instantiation and Evaluation[1]

The approach introduced was instantiated as demonstrator artefact by Martin (2016). The demonstrator is called ICEBERG toolkit and is a Java-based implementation. In consists of a JavaFX user interface, a server component running in a Java container and a repository representing a triple store based on Jena. ICEBERG was extended with a process execution element, which is implemented in the demonstrator using the Camunda BPM framework. Further technical details can be found in Martin (2016). The approach was evaluated in a triangulated evaluation setting. The first and most comprehensive evaluation is the summative evaluation presented in Sects. 8.1 and 8.2. In addition to the summative evaluation, the approach was evaluated in two

---

[1] Results and some verbatim passages in this following section have been taken from the first author's Ph.D. thesis (Martin 2016).

further contexts as described by Martin et al. (2017) and Thönssen et al. (2016), which provide additional evidence and confirmation of the validity.

8.1 Summative Evaluation Using the Procedure Model

The summative evaluation of the procedure model introduced in Sect. 7 uses the application scenario as described in Sect. 3 (Martin 2016). Figure 8 shows the selection of similar cases for a new application of the admission process. The ICEBERG toolkit app is linked to a process execution component, a workflow engine. The cases that can be retrieved consist of case content including process fragments modeled in BPFM comprising procedural knowledge as depicted in Fig. 5.

8.2 Summative Evaluation of Similarity Model

This section describes an evaluation of the similarity performance of the ICEBERG-PE approach using the application scenario described in Sect. 3. This

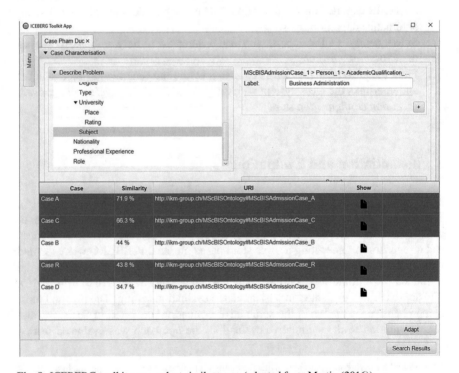

Fig. 8 ICEBERG toolkit app—select similar cases (adapted from Martin (2016))

evaluation was carried out with two stakeholders of the admission process (Martin 2016). Expert A and expert B are interviewers who are part of the MSc BIS interview team performing the "validate eligibility" interviews to validate a candidate, as depicted in Fig. 2. The procedure of the evaluation was as follows. First, the experts were asked to create a similarity configuration. Then they had to rank the cases based on two query cases in task 1. They then had to rank randomly-selected cases according to two randomly-selected query cases in task 2. Finally, the prototype was used to run the retrieval with the same cases in the repository and the same query cases. Tables 1 and 2 show the results of the evaluation. On the left side are the results of expert A's two query cases and on the right, the results of expert B's. In the expectation columns, the experts assigned a number, which represents the similarity of the repository case to the query case. The result columns show the similarity result of the retrieval system as a percentage.

In task one, the experts were asked to rank four cases according to the similarity of two query cases. The experts were asked to assign numbers between 1 (highest similarity) and 4 (lowest similarity) to the cases. The results of task 1 as depicted in Table 1 were as expected. Both experts reported that the comparison of four cases is manageable; more cases would be unmanageable. According to the experts, using a system such as ICEBERG to compare more cases would improve the retrieval significantly.

In task two, the experts were exposed to a more challenging task. They had to randomly select 6 repository cases and 2 query cases out of 66 anonymized cases. The experts were then asked to assign numbers (expectation) between 1 (highest similarity) and 6 (lowest similarity) to the cases. As depicted in Table 2, the results of expert A were almost accurate, and one result of expert B was almost satisfactory. Neverthe-

Table 1 Evaluation task 1 with similarity model of expert A and B (adapted from Martin (2016))

Expert A	Query case Q1		Query case Q2		Expert B	Query case Q1		Query case Q2	
Repository case	Expectation	Result (%)	Expectation	Result (%)	Repository case	Expectation	Result (%)	Expectation	Result (%)
A	4	41	1	67	A	3	44	1	72
B	1	55	3	44	B	2	61	3	44
C	3	40	2	60	C	4	53	2	66
D	1	55	4	35	D	1	61	4	35

Table 2 Evaluation task 2 with similarity model of expert A and B (adapted from Martin (2016))

Expert A	Query case 42		Query case 17		Expert B	Query case 59		Query case 64	
Repository case	Expectation	Result (%)	Expectation	Result (%)	Repository case	Expectation	Result (%)	Expectation	Result (%)
9	5	63	2	54	12	5	53	6	52
24	6	54	1	73	54	2	43	3	54
35	1	72	3	62	7	3	45	2	49
39	2	74	5	57	32	1	47	1	65
56	3	65	6	53	15	6	55	4	62
65	4	64	4	60	41	4	44	5	50

less, a final discussion with the experts revealed some difficulties and suggestions for future research. Both experts stated that it was demanding and difficult to compare more than four cases at once. Although the results may not have been perfect, due to the number of repository cases, the experts attested and considered the approach a significant facilitation of the admission process and the corresponding knowledge work. According to expert B, it would be worthwhile to investigate the possibility of adjusting the similarity configuration model instantly in future research. For specific and unusual cases, it would make sense to weight certain characterization elements differently. The experts suggested that the similarity configuration should be visible, in order to be able to adjust the weights.

9   Discussion and Conclusion

The approach presented in this chapter addresses the research problem of identifying structured elements or process fragments from knowledge work and reusing them with a process execution scenario.

This chapter *contributes to the body of knowledge* on CBR concerning a new approach to support the execution of knowledge-intensive business processes by integrating CBR. This approach was implemented as a reusable open-source proto-type for running experiments and was tested using a real-world application scenario.

Moreover, this research work considers the real-world context right from the beginning, which ensures that the results *contribute to business practice*.

Future work in this area could be as follows: An individual similarity pre-configuration where stakeholders can create a viewpoint model on an individual basis would help them to adapt the CBR to their specific needs. Likewise, an adjustable sim-ilarity configuration where users could carry out the similarity configuration before executing a retrieval step would provide further results that are accurate as well. In both cases, this would require a further cognitively adequate representation of the similarity mechanism, which leads to the final recommendation for future work of a visualization of the similarity configuration.

References

Adams M, Edmond D, ter Hofstede AHM (2003) The application of activity theory to dynamic workflow adaptation issues. In: Proceedings of the 2003 Pacific Asia conference on information systems (PACIS 2003), pp 1836–1852

Allemang D, Hendler JA (2008) Semantic web for the working ontologist: effective modeling in RDFS and OWL. Morgan Kaufmann/Elsevier, San Francisco

Bach K, Althoff K-D (2012) Developing Case-based reasoning applications using myCBR 3. In: Agudo B, Watson I (eds) Case-based reasoning research and development SE—4, vol 7466. Springer, Heidelberg, pp 17–31. https://doi.org/10.1007/978-3-642-32986-9_4

Bello-Tomás JJ, González-Calero PA, Díaz-Agudo B (2004) JColibri: An object-oriented framework for building CBR systems. In: Funk P, González Calero P (eds) Advances in case-based reasoning SE—4, vol 3155. Springer, Heidelberg, pp 32–46. https://doi.org/10.1007/978-3-540-28631-8_4

Bergmann R (2002) Experience management: foundations, development methodology, and internet-based applications. Bergmann R (ed) vol 2432. Springer, Heidelberg. https://doi.org/10.1007/3-540-45759-3

Bergmann R, Althoff K-DK, Minor M, Reichle M, Bach K (2009) Case-based reasoning—introduction and recent developments. Künstliche Intelligenz: Spec Issue Case-Based Reasoning 23(1):5–11

Bergmann R, Freßmann A, Maximini K, Maximini R, Sauer T (2006) Case-based support for collaborative business. In: Roth-Berghofer T, Göker M, Güvenir HA (eds), Proceedings of the 8th European Conference, ECCBR 2006 Fethiye, Turkey, 4–7 Sept 2006, vol 4106. Springer, Heidelberg, pp 519–533. https://doi.org/10.1007/11805816_38

Bergmann R, Schaaf M (2003) Structural case-based reasoning and ontology-based knowledge management: a perfect match? J Univ Comput Sci 9(7):608–626. http://www.jucs.org/jucs_9_7/structural_case_based_reasoning/bergmann_r.pdf. Accessed 19 April 2013

Bichindaritz I (2004) Mémoire: case based reasoning meets the semantic web in biology and medicine. In: Funk P, González Calero P (eds) Advances in case-based reasoning SE—5, vol 3155. Springer, Heidelberg, pp 47–61. https://doi.org/10.1007/978-3-540-28631-8_5

Brander S, Hinkelmann K, Hu B, Martin A, Riss U, Thönssen B, Witschel H (2011) Refining process models through the analysis of informal work practice

Cognini R, Corradini F, Polini A, Re B (2015a) Using data-object flow relations to derive control flow variants in configurable business processes. In: Fournier F, Mendling J (eds), International workshops business process management workshops: BPM 2014, Eindhoven, The Netherlands, 7–8 Sept 2014, Revised Papers. Springer International Publishing, Cham, pp 210–221. https://doi.org/10.1007/978-3-319-15895-2_19

Cognini R, Corradini F, Polini A, Re B (2015b) Extending feature models to express variability in business process models. In: Persson A, Stirna J (eds), Proceedings of the international workshops advanced information systems engineering workshops: CAiSE 2015, Stockholm, Sweden, 8–9 June 2015, Springer International Publishing, Cham, pp 245–256. https://doi.org/10.1007/978-3-319-19243-7_24

Cognini R, Hinkelmann K, Martin A (2016) A case modelling language for process variant management in case-based reasoning. In: Reichert M, Reijers AH (eds), 13th international workshops business process management workshops: BPM 2015, 4th international workshop on adaptive case management and other non-workflow approaches to BPM: AdaptiveCM 2015, Innsbruck, Austria, 31 Aug–3 Sept 2015, Revised. Springer International Publishing, Cham, pp 30–42. https://doi.org/10.1007/978-3-319-42887-1_3

Cohen D, Crabtree B (2006) Triangulation. Qualitative research guidelines project, Robert Wood Johnson Foundation. http://www.qualres.org/HomeTria-3692.html. Accessed 26 May 2016

Davenport TH (2005) Knowledge work processes. In: Thinking for a living: how to get better performance and results from knowledge workers

Davenport TH (2010) Process management for knowledge work. In: vom Brocke J, Rosemann M (eds) Handbook on business process management 1. Springer, Heidelberg, pp 17–36. https://doi. org/10.1007/978-3-642-00416-2_2

Eberle H, Unger T, Leymann F (2009) Process fragments. In: Lecture notes in computer science (including subseries lecture notes in artificial intelligence and lecture notes in bioinformatics), LNCS, vol 5870, pp 398–405. https://doi.org/10.1007/978-3-642-05148-7_29

El-Farr HK (2009) Knowledge work and workers: a critical literature review, vol 1

Gadatsch A (2012) Grundkurs Geschäftsprozess-Management (7. Auflage.). Wiesbaden: Vieweg+Teubner Verlag. https://doi.org/10.1007/978-3-8348-2428-8

Hevner AR, March ST, Park J, Ram S (2004) Design science in information systems research. MIS Quarterly 28(1):75–105. http://www.jstor.org/stable/25148625. Accessed 13 Apr 2011

Ju K, Su B, Zhou D, Zhang Y (2016) An incentive-oriented early warning system for predicting the co-movements between oil price shocks and macroeconomy. Appl Energy 163:452–463. https://doi.org/10.1016/j.apenergy.2015.11.015

Kolodner JL (1993) Case-based reasoning. Morgan Kaufmann Publishers, San Mateo, CA

Leake DB (1996) CBR in context: the present and future. In: Leake DB (ed) Case-based reasoning: experiences, lessons, and future directions. AAAI Press/MIT Press, Menlo Park, pp 1–35

Madhusudan T, Zhao JL, Marshall B (2004) A case-based reasoning framework for workflow model management. Data Knowl Eng 50(1):87–115. http://www.sciencedirect.com/science/article/pii/S0169023X04000060

Martin A (2016) A combined case-based reasoning and process execution approach for knowledge-intensive work. PhD thesis. University of South Africa, Pretoria. Retrieved from http://hdl.handle.net/10500/22796

Martin A, Brun R (2010) Agile process execution with KISSmir. In: 5th international workshop on semantic business process management collocated with 7th extended semantic web conference. Heraklion, Greece

Martin A, Emmenegger S, Hinkelmann K, Thönssen B (2017) A viewpoint-based case-based reasoning approach utilising an enterprise architecture ontology for experience management. Enterp Inf Syst 11(4):551–575. https://doi.org/10.1080/17517575.2016.1161239

Martin A, Emmenegger S, Wilke G (2013) Integrating an enterprise architecture ontology in a case-based reasoning approach for project knowledge. In: Proceedings of the first international conference on enterprise systems: ES 2013. IEEE, Cape Town, pp. 1–12. https://doi.org/10.1109/es.2013.6690082

Maximini K, Maximini R (2007) Collaborative agent-based knowledge engine. wi2.uni-trier.de. Trier, Germany. http://www.wi2.uni-trier.de/publications/2007_CAKE_Maximini.pdf. Accessed 9 Aug 2012

Patton MQ (2015) Qualitative research and evaluation methods: integrating theory and practice, 4th ed. SAGE Publications Inc

Peffers K, Tuunanen T, Rothenberger MA, Chatterjee S (2007) A design science research methodology for information systems research. J Manage Inf Syst 24(3):45–77. https://doi.org/10.2753/MIS0742-1222240302

Recio-García JA, Díaz-Agudo B, González-Calero P (2008) jCOLIBRI2 Tutorial. Department of Software Engineering and Artificial Intelligence, University Complutense of Madrid, Madrid

Recio-García JA, González-Calero PA, Díaz-Agudo B (2014) jcolibri2: a framework for building case-based reasoning systems. Sci Comput Prog 79(Group 910494):126–145. https://doi.org/10.1016/j.scico.2012.04.002

Roth-Berghofer T, Bahls D (2008) Explanation capabilities of the open source case-based reasoning tool myCBR. In: Petridis M, Wiratunga N (eds), UK workshop on case-based reasoning UKCBR 2008, pp 23–34. http://mycbr-project.net/downloads/ukcbr08.pdf. Accessed 29 Mar 2013

Thönssen B, Witschel H-F, Hinkelmann K, Martin A (2016) Experience knowledge mechanisms and representation. Learn PAd—Model-Based Social Learning for Public Administrations—EU FP7-ICT-2013-11/619583, Pisa, Italy. puma.isti.cnr.it/dfdownloadnew.php?ident=/LPAd/2016-EC-005

Vaishnavi V, Kuechler B (2004) Design science research in information systems. Page on design science research in information systems (IS); last updated: 15 Nov 2015. http://www.desrist.org/design-research-in-information-systems/. Accessed 15 Nov 2015

van Heijst G, Schreiber AT, Wielinga B (1997) Using explicit ontologies in KBS development. Int J Hum Comput Stud 46(2–3):183–292. https://doi.org/10.1006/ijhc.1996.0090

W3C (2012) OWL 2 web ontology language. http://www.w3.org/TR/owl2-overview/

W3C (2014a) RDF 1.1 concepts and abstract syntax. http://www.w3.org/TR/rdf11-concepts/

W3C (2014b) RDF Schema 1.1. http://www.w3.org/TR/rdf-schema/

W3C (2014c) RDF 1.1 Turtle. http://www.w3.org/TR/turtle/

Watson I (1999) Case-based reasoning is a methodology not a technology. Knowl-Based Syst 12(5–6):303–308. https://doi.org/10.1016/S0950-7051(99)00020-9

Witschel H-F, Martin A, Emmenegger S, Lutz J (2015) A new retrieval function for ontology-based complex case descriptions. In: International workshop case-based reasoning CBR-MD 2015. ibai-publishing, Hamburg

Road to Agile Requirements Engineering: Lessons Learned from a Web App Project

Rainer Telesko

Abstract This chapter describes the research project Companion conducted at the University of Applied Sciences and Arts Northwestern Switzerland FHNW and its relationship to the research area Agile Requirements Engineering (ARE). ARE aims to establish requirements engineering practices, which are customized for agile development methodologies such like Scrum, XP, etc. Within the Companion project a Web App was developed to promote mental health of adolescents taking their first steps into working life. The description starts by giving some key information about the project, its context, the development of the Web App and the weaknesses observed during software engineering activities. Next, classical requirements engineering and ARE are compared and specific challenges for ARE based on experiences from the industry are presented. For the challenges in ARE, specific practices have been proposed which are described together with their limitations. This chapter ends with a selection of adequate ARE practices which may increase the performance of software engineering in situations similar to the context of the Companion project.

Keywords Agile requirements engineering · Software development · Scrum Companion app

1 The Companion Project: An Overview

The Companion project was funded by the Commission for Technology and Innovation CTI in Switzerland and conducted from March 2013 to December 2014. Project partners were Health Promotion Switzerland, the School of Applied Psychology at the Zurich University of Applied Sciences, and the Institute for Information Systems at the University of Applied Sciences and Arts Northwestern Switzerland.

R. Telesko (✉)
Institute for Information Systems, University of Applied Sciences and Arts Northwestern
Switzerland, Peter Merian-Strasse 86, 4002 Basel, Switzerland
e-mail: rainer.telesko@fhnw.ch

© Springer International Publishing AG 2018
R. Dornberger (ed.), *Business Information Systems and Technology 4.0*,
Studies in Systems, Decision and Control 141,
https://doi.org/10.1007/978-3-319-74322-6_5

Pilot partners for the evaluation of the app were two large Swiss organizations with around 600 participating apprentices.

The main motivation (Bohleber et al. 2016) for the project was the fact that Swiss adolescents starting their working life with an apprenticeship are beginning an adult life work rhythm at a young age for which they are not prepared. Stressful situations may occur, for example when adolescents face completely new learning content in school, are not able to master both school and work to a sufficient level, have problems keeping up with the working rhythm and pace, etc. Up to now, no specific mental health programs have been implemented to support these young people in their daily challenges. The Companion project aims to fill exactly this gap.

Based on the fact that smartphones are a core part of the life of most adolescents, a web app was developed to foster mental health in adolescents taking their first steps into working life. One core hypothesis (Bohleber et al. 2016) behind the solution was that enhanced social support among peers is a cornerstone in promoting well-being and reducing stress levels.

The core features of the Companion app (Bohleber et al. 2016; Telesko and Bendel 2017) were the following:

- A peer mentoring system: Peer mentoring was implemented by apprentices asking other users to become their mentor and to provide support in daily life (e.g. school, social situations, etc.).
- An individual profile for each user, messaging and discussion groups: These features are very similar to social media platforms such as Facebook, Xing, etc.
- Links to interactive and informative websites on mental health-related issues (e.g. psychological tests, sexuality, drug use).
- Links to dedicated websites concerning leisure activities for adolescents.
- An anonymous professional counseling service.

The Companion app was implemented as a web app (i.e. not a native solution) and could be run on a smartphone, tablet or desktop computer.

One issue which is critical in the context of social media applications is cyber mobbing. In order to avoid negative effects when using the app, an imprint, terms of business and a netiquette were developed and presented to the apprentices (Telesko and Bendel 2017).

The project itself can be separated into three phases (Telesko and Bendel 2017): a planning phase, a pilot phase and a dissemination phase.

- The planning phase: In the planning phase the project team was formed and initial ideas about the app and its features were developed. Also the concept for peer mentoring was also outlined.
- The pilot phase with evaluation: In this phase the requirements were elicited and a draft version of the app was developed. Feedback from apprentices (e.g. regarding user interface, additional features) helped to continuously improve the quality. The two pilot partners tested the app during a nine-month pilot phase. The project team set up a statistical experiment in order to find out whether usage of the app significantly contributed to reducing stress. The experiment was realized by setting

up a random intervention and control group. The result of the experiment was that the intervention via the Companion app had no significant impact on chronic stress levels or the perception of social support (Bohleber et al. 2016).

- Dissemination phase: In this phase the results of the project were presented in publications and at scientific events and additional pilot partners were sought to test future releases of the app. The project partners are currently seeking customers who are interested in contributing new ideas and developing Companion further, also for new scenarios. Responsibility for the development and maintenance was handed over to a professional IT company.

2 The Companion Project: Evaluating Software Engineering Activities

This section describes the software engineering activities conducted in the Companion project and identifies weaknesses impacting the quality of the outcome (von Wyl et al. 2014).

2.1 Development Methodology

The project team decided to choose Scrum (Schwaber and Sutherland 2012) as the development methodology. The rationale for this decision was that the implementation partner already had experience with Scum in research projects and the nature of the project was agile, because there was no stable, committed view about the requirements to be implemented at the outset.

The concrete usage of Scrum had several shortfalls. First, there were very often no formal reviews and retrospectives at the end of the Sprints which could provide insight into the details and quality of the implementation. Furthermore, the role of the Scrum Master, who should have acted as the process owner for the development, was only used sporadically. This had an impact on the quality of the communication and cooperation between the technical team and the other project members.

2.2 Requirements Engineering

The elicitation of requirements was initiated in a plenary workshop. The gathered requirements, which were typically minimal for a Scrum project, were documented in an Excel-based product backlog, analyzed and prioritized and served as a basis for the derivation of the Sprint backlogs. The requirements collected were discussed with the apprentices during the information meetings at the beginning of the project. Overall, this phase was successful. However, collaboration between the project team and the apprentices should have been intensified in order to derive more dedicated features with added value in promoting mental health. The fact that the Companion app contains a relatively large number of features of "classical" social media applications (e.g. setting up profiles, groups, messaging, etc.) such as Facebook, Instagram, Xing, etc. turned out to be an obstacle to promoting the unique selling proposition.

2.3   Implementation

Based on the fact that multiple operating systems had to be supported and no smartphone-specific functionality was needed, the decision within the project team was to go for a web app and not a native app. The main advantage of this decision was that the scarce resources could be efficiently used. The main disadvantage as became evident during the evaluation was that apprentices were missing the typical "app feeling".

The Play 2 framework (Richard-Foy 2014) was selected for the development, allowing the development of Java- and Scala-based solutions with support for well-known libraries and routine tasks, while requiring very little server infrastructure.

During the implementation phase it became clear that there is a goal conflict between a prototypical software product created in a research project with limited resources and a commercial product with a high-performing user interface, as was expected by the apprentices during the evaluation. An advanced prototype developed in a research project can never be a real competitor to high quality software, especially when we examine the most important criteria: bug-free software through extensive testing, a sophisticated user interface using cutting-edge technology and the consideration of critical non-functional requirement categories such as usability, reliability and stability.

2.4   Testing

Testing activities are always crucial in software engineering projects because of time pressure. In particular in agile projects where little knowledge about the features to be implemented is available at the very beginning, testing activities are radically shortened. This was the case for the Companion project and led to the situation that a fully-fledged version of Companion could not initially be presented to some of the apprentices in one pilot run.

Using appropriate tools helped to save time during development. The Play 2 framework supported the quick development of routine features (e.g. login, building forms, etc.) and therefore enabled the programmers to focus on complex features such as messaging or group functionality. The open-source collaboration platform Gitlab (https://gitlab.com) was used extensively in the project team to report issues and bugs and allowed the developers to quickly build a new increment. The start of the two pilots took place around six months after the project start and therefore quite early in the project. This led to the fact that testing activities were very often postponed and conducted at the end of the development in "extra" Sprints reserved for testing.

3 Agile Requirements Engineering (ARE)

Requirements engineering is the discipline which defines how to collect, document, check and manage requirements (Rupp and Pohl 2015). In the "classical" world, when using the waterfall model, requirements are described completely and in-detail before starting work on the design.

In the agile paradigm the approach is different. Basically, there is no real requirements engineering. Based on the business value of the requirements, they are mostly described solely as user stories at the beginning of the Sprints. A definition reflecting the main idea of ARE is provided by Heikkilä et al. (2015, p. 205):

> In agile RE, the requirements are elicited, analyzed and specified in an ongoing and close collaboration with a customer or customer representative in order to achieve high reactivity to changes in the requirements and in the environment. Continuous requirements re-evaluation is vital for the success of the solution system, and the close collaboration with the customer or customer representative is the essential method of requirements and system validation.

The ARE discipline is now investigating to what extent this abstract definition can be concretized to optimize the performance of development projects. Basically, there are two completely different viewpoints:

- The first approach is based on the fact that requirements engineering is mostly realized by communication with the stakeholders. The documentation should be—independent of the nature of the project—in any case minimal; classical requirements engineering methods (Oestereich 2010a) such as stakeholder analysis, linguistic techniques like text templates, business process modelling, UML models, etc. should be avoided because they only create overhead.
- The second approach which is investigated further in this chapter aims to find out how classical requirements engineering techniques can be beneficially combined with agile development techniques. The main advantage (Oestereich 2010a) of this approach is that circumstances in a project can be assessed much more quickly than when using only minimal requirements engineering. The added value is therefore the time factor. In a Scrum project the stakeholders have to wait for the review of a Sprint to give their feedback and to possibly change project requirements and features. If more sophisticated techniques are used during the definition of the Sprint backlog, possible obstacles and impediments to implementation can be found much faster. For example, the use of a text template to specify a requirement helps to identify linguistic transformations that are often overlooked when solely

defining user stories, and the usage of an UML activity diagram helps to clarify the complexity of a feature for all team members.

The main question now is what to "invest" in detail for requirements engineering in order to end up with a superior cost benefit ratio. A possible solution could be to organize the techniques in a workbench (with predefined costs for training and use) where one can pick up a technique as needed. Some guiding questions which help to assess "how much" requirements engineering is needed are (Oestereich 2010b):

- How familiar are the stakeholders (i.e. product owner, domain experts, developers, testers, etc.) with the domain?
- How homogeneously are the business processes and use cases executed by different users?
- How many different variants and exceptions have to be considered?
- How new are the procedures in the software compared to the old solution?
- How many different persons are providing or influencing requirements?
- How big is the risk that using very little requirements engineering may lead to overlooked dependencies or details that may later cause extra costs or delays?
- How serious are the consequences if important functional variants or details are forgotten?

Basically, all requirements engineering practices can be used in an agile development context (Oestereich 2010a). The selection of adequate practices for later use in the project can now be made by defining criteria. One criterion might be the effort to learn a practice, e.g. the time and related costs of getting acquainted with basics of UML modeling; another criterion might be the level of existing knowledge about the practices.

In the further course of this chapter we will consider the challenges of agile development as a criterion. The nature of agile development models such as Scrum, Extreme Programming (XP), Feature Driven Development (FDD), etc., raises specific challenges that can be met by the proper use of requirements engineering practices. With such a procedure, we get a superset of beneficial practices that can be further narrowed down by taking into account criteria such as cost, learnability, paper-based or IT-based support, etc.

4 The Procedure Model Scrum

The majority of agile projects are realized with Scrum (Schwaber and Sutherland 2012). Scrum is based on the principles of the Agile Manifesto (http://agilemanifesto. org/) which are

- individuals and interactions over processes and tools;
- working software over comprehensive documentation;
- customer collaboration over contract negotiation;
- responding to change over following a plan.

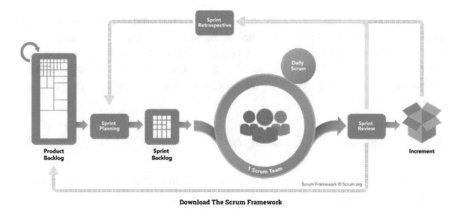

Download The Scrum Framework

Fig. 1 The scrum framework (*Source* https://www.Scrum.org/resources/what-is-Scrum)

Scrum is a lightweight procedure model for the development of software in an iterative and incremental way based on business needs. The increments are called Sprints and typically take three to four weeks. After a Sprint, a running increment of the software is available. Scrum differs from heavyweight procedure models such as the waterfall model, V-model, Rational Unified Process (RUP), not by the usage of specific development techniques, but rather, by incorporating and accepting change as the key driver.

Web applications very often perfectly fit the agile philosophy. Implementation is started with the requirements that provide the greatest business value. The increments produced in the Sprints are rapidly communicated to the customer, which allows new versions of the software to be found quickly.

In Fig. 1 the main elements of the Scrum Framework are depicted and subsequently described (Sutherland and Schwaber 2016).

The Scrum team, which is self-organizing, consists of the development team, the Product Owner (PO) and the Scrum Master. The development team designs, implements, and tests the software during the Sprints. A Definition of Done (DoD) describes in detail what has to be accomplished. The PO defines the overall vision and is responsible for maximizing the value of the product and managing the Product Backlog. The Scrum Master is the "process owner" for Scrum and the contact person if there are for example impediments that seriously endanger the success of a Sprint.

The Product Backlog is an ordered list of everything (e.g. requirements, features, etc.) which might be needed in the project. Based on the business value of each requirement the Scrum team defines the Backlog of each Sprint during the Sprint Planning, i.e. the items selected for the Sprint and a plan to realize the Sprint Goal. The effort for a Sprint may be calculated, for example, via the Sprint Planning Poker, where for each feature each team member plays a numbered card face-down on the table with an estimation for the hours needed. During the Sprint the development team is protected, this means requirements cannot be changed or deleted. The Daily Scrum is held every day in order to synchronize activities, discuss impediments and plan for the next 24 h. At the end of each Sprint the increment solution is presented by the developers with the DoD as a guideline. During this Sprint Review, the Scrum team may adapt the Product Backlog if necessary and thus react to changing requirements or new project circumstances. Finally, the Sprint Retrospective is an opportunity to review the past Sprint and optimize the process for the next Sprint.

5 Overview of ARE Challenges

This section provides an overview of the most important Agile Requirements Engineering challenges.

5.1 Minimal Requirements Documentation

The most important issue is the minimal requirements documentation (Inayat et al. 2015; Elghariani and Kama 2016), which directly emerges from the agile philosophy to promote communication over documentation. User requirements are usually only described in a feature list, user stories or Product/Sprint Backlogs. Their usage in follow-up projects is limited. User stories are not well suited to describing the software design and support traceability, i.e. the relationships between requirements, code and testing over time (Heikkilä et al. 2015). Furthermore, the fluctuation of important team members during or after the project endangers the success of the project.

5.2 Ignoring Non-functional Requirements (NFR)

Non-Functional Requirements such as usability, availability, reliability, maintainability, etc. are very often neglected in Scrum projects (Inayat et al. 2015; Elghariani and Kama 2016). The main reason for this is that customer representatives mainly focus on achieving a correct implementation of the features and have no idea about how the software should function after the termination of the project. The development in Sprints also makes it very hard to measure the stable fulfillment of NFRs because of the possible rapid change of features in consecutive Sprints.

5.3 Inappropriate Software Architecture

Changing requirements may also influence the software architecture defined at the beginning (Heikkilä et al. 2015; Elghariani and Kama 2016). Let us assume that at the very beginning a specific architectural style such as a monolithic, a component-based, a three-tier or a service-oriented architecture was chosen. This decision was made with insufficient knowledge. Later on, when the team sees much more clearly which features are essential, where they should run (on the client or on the server) and how the modules interact, the decision initially taken is mostly irreversible. Furthermore—similar to NFRs—the customer has very often little experience in this matter.

5.4   Insufficient Customer Availability

The availability of the customer throughout the whole project is one of the corner-stones of the agile methods (Inayat et al. 2015; Heikkilä et al. 2015; Elghariani and Kama 2016). The customer should contribute to the specification of the requirements and give feedback during the Sprint Reviews. However, this is very often not possible due to budget and time constraints on the customer side. Most of the customers expect that all the relevant work will be achieved without their involvement and also express this attitude when budgeting IT-projects. The absence of the customer now implies that the direction of the project will somehow become unclear and shifts the nature of the project towards classical development where the acceptance test is carried out after finishing the complete implementation.

5.5   Imprecise Effort Estimation

Due to unstable requirements, making precise upfront effort estimations in a Scrum project is almost impossible (Inayat et al. 2015; Heikkilä et al. 2015; Elghariani and Kama 2016). This issue is especially important when the project partners are thinking about setting up a fixed-price fixed-scope contract.

6   Proposed Solutions for ARE Challenges

The ARE challenges listed in the previous section do not include problems that emerge from an incorrect application of Scrum itself. For example, an "incorrect requirements prioritization" discussed in (Inayat et al. 2015) will cause problems later on because the project team uses an inadequate algorithm for prioritization, defines incorrect weights, and is not aware of important stakeholders, etc.

There are dependencies between the challenges, which make it easier to identify and select possible solutions with limited effort. For example, the "minimal requirements documentation" contributes to causing the "inappropriate software architecture" and "imprecise effort estimation" challenges because user stories are in most cases not an adequate means to describe software architecture and to estimate the

Table 1 Proposed solutions for ARE challenges

Challenge number	Challenge description	Influencing other challenges	Proposed solutions
1	Minimal requirements documentation	3, 5	Enhanced documentation, prototyping, test-driven development (TDD)
2	Ignoring non-functional requirements (NFR)	3, 5	User stories for NFR, visual modeling of NFR
3	Inappropriate software architecture	–	Test-driven development (TDD)
4	Insufficient customer availability	1, 5	Proxy customer, dedicated Requirements Engineer
5	Imprecise effort estimation	–	–

effort for the project. Table 1 now contains some proposed solutions for the challenges together with their dependencies. Most of the proposed solutions are described afterwards.

6.1 Proposed Solutions for Minimal Requirements Documentation

An extension to a minimal documentation is useful as long as it complies with the philosophy of Scrum. Two examples of such an extension are given in (Leutbecher 2012):

- T. Weilkiens proposes extending a user story with elements of a use case description. Activity, state chart diagrams and scenarios (normal flow, alternative flow, and exceptional flow) can be used here.
- B. Oestereich proposes extending the Product Backlog description with additional elements such as identifier, name/title, short description, acceptance criteria.

A different approach is to use prototyping (Käpyaho and Kauppinen 2015) in order to provide the customer with an idea about the system as early as possible. Using test-driven development (TDD), which is described in detail in challenge 3, may also be an alternative to "classical" documentation like a user story or a use case. The advantages of this technique, which is especially popular among developers, can be best expressed with the citation "You write code that talks about what the system's behavior should be. So you end up writing very explicit specifications and not 'tests'" (Cao and Ramesh 2008).

6.2   Proposed Solutions for Ignoring Non-functional Requirements (NFRs)

The first approach is to use the well-known text template for user stories (Rupp and Pohl 2015) which helps reduce linguistic transformational effects:

- Using the text template for a functional requirement: "The system shall provide the user with the ability to back up all data."
- Using the text template for an NFR: "The library system shall complete each saving process within a maximum duration of one second."

The main advantage is that it is very easy to learn and customers can also contribute to the specification. The approach described in Farid and Mitropoulos (2012) goes one step further. They propose NORMATIC as an approach to adequately identify, model, and link NFRs and their potential solutions with functional requirements. For this purpose they propose an agile requirements taxonomy representing

- Functional requirements (Agile Use Cases—AUC)
- NFRs (Agile Loose Cases—ALC) and
- Agile Choose Cases (ACC) for the potential solutions of NFRs.

The NORMAP methodology is supported by a semi-automatic tool, which allows the visual modelling of AUC, ALC and ACC and the management of metadata for requirements such as quality attributes (ambiguous, validated, etc.), assigned Sprint and release, risk score, etc.

6.3   Proposed Solutions for an Inappropriate Software Architecture

One possible solution might be the usage of test-driven development (TDD) which is also reported as widely accepted ARE practice in industry (Cao and Ramesh 2008).

TDD is a software design technique whereby tests are created before writing new functional code. It is mainly applied with the "Red-Green-Refactor"-cycle:

- Step 1: Write a test that fails.
- Step 2: Write just enough code to pass the test.
- Step 3: Refactor: Optimize and clean up the code without adding new functionality.

This procedure ensures that all functionality is tested and only code is written that is needed. An extension is now to also apply TDD early in Scrum to NFRs such as performance, scalability, portability, etc. If such tests fail this is an "early warning signal" that the existing software architecture might not be appropriate.

6.4 Proposed Solutions for Insufficient Customer Availability

One solution proposed in the literature is to nominate the PO as proxy customer (Inayat et al. 2015; Elghariani and Kama 2016). One has to be very cautious here because the PO in a Scrum project "is responsible for maximizing the value of the product" and "the sole person responsible for managing the Product Backlog" (Sutherland and Schwaber 2016). Assigning the tasks of a customer to the PO might cause an overload and also create goal conflicts.

A different solution is proposed in Hochmüller (2011) where the author proposes introducing the new role of a requirements engineer in a Scrum team. This requirements engineer should act as a "customer companion" when accompanying the customer to the site of the developers. The main tasks of this new role consist of providing support with the effort estimation, identifying stakeholders, writing and validating requirements. The main benefits of this approach are that the "customer companion" will contribute to increasing the quality level of overall requirements engineering and the developers can concentrate on their work and reduce the number of tedious customer inquiries. The main disadvantage is the additional costs created by this new role, which should, hopefully, be compensated by the reduced costs of poor quality of the software.

6.5 Proposed Solutions for Imprecise Effort Estimations

This issue seems to be the most challenging because also in Heikkilä et al. (2015) it is reported that there was no solution found in the scientific community. Basically, only an extensive requirements analysis can improve estimations, which is an inherent contradiction of the nature of Scrum.

While fixed-price, variable-scope and variable-price fixed scope projects work well with Scrum, fixed-price, fixed-scope projects should be better realized with a non-agile model.

7 Lessons Learned from the Companion Project

With regard to reaching the project goals, challenges 1 (minimal requirements documentation), 2 (ignoring non-functional requirements) and 4 (insufficient customer availability) were the most problematic.

The minimal requirements documentation made it almost impossible for programmers to join the project later and to carry out the project based on well-defined releases. This also became evident when the source code was handed over to a professional IT company after termination of the project. Ignoring or neglecting non-functional requirements is typical for a research project where the main emphasis is on functional requirements. The specific setting of the Companion project with developing and evaluating a state-of-the-art web app makes NFR even more important and requires specific attention to proper balancing research and design issues. The insufficient integration of the pilot partners and apprentices during requirements elicitation was one of the reasons why the aspired usage of the software was not achieved. In the Internet age it is absolutely necessary to continuously hear the "Voice of the Customer" and to develop a product in close collaboration with all stakeholders.

In retrospect we considered that the following would have been adequate techniques to avoid such issues:

- For challenge 1, the extension of Product Backlog entries with use case elements and performing prototyping is helpful. In particular, prototyping, when used in a pre-study with pre-selected apprentices, reveals which features to focus on in the regular project.
- For challenge 2, two solutions seem adequate. The first solution is to describe NFR with extra user stories. This also helps to create awareness for this topic during development. The second solution is to add a new project partner—a web design company—which is responsible for all user interface and user experience issues.
- For challenge 4, a dedicated requirements engineer can assure the necessary level of documentation and at the same moment unburden the developers so that they can concentrate on software development.

8   Conclusions and Outlook

In this chapter, we investigated how requirements engineering may be embedded into agile development. Based on the nature of agile development methods such as Scrum, some challenges and possible practices were identified. Future research should focus on empirical results when applying ARE practices and especially to what extent using a certain practice contributes to achieving the project goals with acceptable costs. With this knowledge the industry can then capitalize on an ARE workbench containing proven requirements engineering practices with a superior cost benefit ratio.

References

Bohleber L, Crameri A, Eich-Stierli B, Telesko R, von Wyl A (2016) Can we foster a culture of peer support and promote mental health in adolescence using a web-based app? A control group study. JMIR Ment Health 3:e45. https://doi.org/10.2196/mental.5597

Cao L, Ramesh B (2008) Agile requirements engineering practices: an empirical study. IEEE Softw 25:60–67. https://doi.org/10.1109/MS.2008.1

Elghariani K, Kama N (2016) Review on agile requirements engineering challenges. In: 2016 3rd International conference on computer and information sciences (ICCOINS), pp 507–512

Farid WM, Mitropoulos FJ (2012) NORMATIC: a visual tool for modeling non-functional requirements in agile processes. In: 2012 Proceedings of IEEE Southeastcon, pp 1–8

Heikkilä VT, Damian D, Lassenius C, Paasivaara M (2015) A mapping study on requirements engineering in agile software development. In: 2015 41st Euromicro conference on software engineering and advanced applications, pp 199–207

Hochmüller E (2011) The requirements engineer as a liaison officer in agile software development. In: Proceedings of the 1st workshop on agile requirements engineering. ACM, New York, NY, USA, pp 2:1–2:4

Inayat I, Moraes L, Daneva M, Salim SS (2015) A reflection on agile requirements engineering: solutions brought and challenges posed. In: Scientific workshop proceedings of the XP2015. ACM Press, Helsinki, pp 1–7

Käpyaho M, Kauppinen M (2015) Agile requirements engineering with prototyping: a case study. In: 2015 IEEE 23rd International requirements engineering conference (RE), pp 334–343

Leutbecher E (2012) Agiles requirements engineering professionalisieren. https://entwicklertag.de/karlsruhe/2012/sites/entwicklertag.de.karlsruhe.2012/files/LeutbecherAgilesREprofessionalisierenV1_00publishFolien.pdf. Accessed 4 Apr 2017

Oestereich B (2010a) Aufgespalten - Agiles Anforderungsmanagement—gibt es das? Heise iX 8:93–95

Oestereich B (2010b) Gedanken über agiles requirements engineering. In: heise developer. http://www.heise.de/developer/artikel/Gedanken-ueber-agiles-Requirements-Engineering-948348.html. Accessed 10 Mar 2017

Richard-Foy J (2014) Play framework essentials. Packt Publishing, Birmingham

Rupp C, Pohl K (2015) Requirements engineering fundamentals: a study guide for the certified professional for requirements engineering exam—foundation level—IREB complaint, 2nd edn. Rocky Nook, Santa Barbara, CA

Schwaber K, Sutherland JV (2012) Software in 30 days: how agile managers beat the odds, delight their customers, and leave competitors in the dust. Wiley, Hoboken, NJ

Sutherland J, Schwaber K (2016) The Scrum guide™. http://www.scrumguides.org/scrum-guide.html. Accessed 13 Mar 2017

Telesko R, Bendel O (2017) Companion: Eine App zur Unterstützung der Peer-Kultur in Betrieben. In: Pfannstiel MA, Da-Cruz P, Mehlich H (eds) Digitale Transformation von Dienstleistungen im Gesundheitswesen I. Springer Fachmedien Wiesbaden, pp 265–275

von Wyl A, Amstad F, Telesko R (2014) Die Companion App: Ein Pilotprojekt zur Förderung der psychischen Gesundheit bei Jugendlichen im betrieblichen Umfeld

Part II
E-Business Applications

E-Business in the Era of Digital Transformation

Uwe Leimstoll, Achim Dannecker, Hanspeter Knechtli, Michael Quade,
Christian Tanner and Ralf Wölfle

Abstract In recent years, the development of information technology has reached
a new level of evolution. It should be noted, however, that digitalization and digital
transformation are not fundamentally new phenomena. A phase marked by strong
digitalization, for example, was the result of the increased business use of the Internet
in the early 1990s. The era of e-business began. The focus was on the digitalization of
cross-company processes, which also brought new business models and interlinked
value-added structures. Against the background of current developments, the question arises as to what role e-business now plays in digital transformation. The aim of
this chapter is to show the current developments in e-business and their relationship to
digital transformation. Following an introductory clarification of concepts and a consideration of the importance of mobile computing, the classic areas of e-business are
addressed, in particular e-commerce, e-procurement and e-organization. The results
show that e-business has always involved digital transformation processes and that
these three perspectives of e-business are still helpful in analyzing the effects of
digital transformation and identifying digitalization potential in and between companies. Besides information technology itself, the concepts of e-business become an
important driver of digital transformation.

Keywords Digital transformation · Mobile computing · E-Business
E-Commerce · E-Procurement · E-Organization

1 Introduction

In the middle of the 2010s, *digitalization* and *digital transformation* are among the
top issues in the context of social and economic developments. It must be remem-
bered that the *Digital Age* and a *New Economy* had already been proclaimed some

U. Leimstoll (✉) · A. Dannecker · H. Knechtli · M. Quade · C. Tanner · R. Wölfle
Institute for Information Systems, University of Applied Sciences and Arts Northwestern
Switzerland, Peter Merian-Strasse 86, 4002 Basel, Switzerland
e-mail: uwe.leimstoll@fhnw.ch

© Springer International Publishing AG 2018
R. Dornberger (ed.), *Business Information Systems and Technology 4.0*,
Studies in Systems, Decision and Control 141,
https://doi.org/10.1007/978-3-319-74322-6_6

twenty years ago (Tapscott 1995). They brought forth concepts such as *e-business* and *e-commerce*, which have lasted to the present day. Digitalization in the narrower understanding of computer-aided, binary data processing goes back even to the 1960s and is generally regarded as the catalyst of the *Third Industrial Revolution*; it is also called the *Digital Revolution*.

Against this background, the question arises as to what exactly lies behind today's very popular concept of digital transformation and how the current form of digitalization differs from the previous stages of development of information technology (IT). In this context, e-business concepts have to be analyzed in order to answer the next question, addressing which current changes may happen in these areas and to what extent they are to be attributed to digital transformation or connected to it.

The purpose of this chapter is to bring clarity to the current conceptual world of digitalization, on the one hand, and on the other, to highlight the particular features of digital transformation. Furthermore, current developments in the sub-areas of e-business shall illustrate how digital transformation is already affecting the economy and society today, and how it can be implemented in companies or other organizations.

The following section first describes the central concepts of digitalization and e-business and associates digital transformation with the evolution of IT. Section 3 describes one of the key drivers of digital transformation: the mobile devices. The following sections deal with current developments in the areas of e-commerce, e-procurement and e-organization. The final section summarizes the results and draws conclusions.

2 Digital Transformation—The Third Evolutionary Stage of IT

This section first shows how the development of IT reached a new, third evolutionary stage at the end of the 2010s. The most important concepts on the topic of digitalization, which have been emerging since about 2010, are defined[1].

The third evolutionary stage of IT was preceded by two stages: In the first three decades of digital information processing, IT primarily served to increase efficiency in existing forms of organization. Information systems were rarely interlinked. The type of products offered on the market and the nature of the interaction with customers and business partners did not change fundamentally. While major advances in productivity have caused significant shifts in the labor market, structural changes have not yet been a major issue. At the beginning of the 1990s, there was a paradigm shift, which initiated the start of the second evolutionary stage: The invention of the World Wide Web marked the beginning of the explosive expansion of internet use:

---

[1]The definitions of the newer concepts are based on the interpretation of current developments in Klaus Schwab's work "The Fourth Industrial Revolution" (2016). A more comprehensive derivation and definition of concepts of digitalization is found in Wölfle (2016).

Incompatible systems were connected, broad, inter-organizational networking and integration began, which created access to a wealth of information and functions. It quickly became apparent that the impact of the Internet cannot be limited to IT itself or to the business world. Already in 1995, Don Tapscott spoke of the *digital economy* and described it as "the economy for the age of Networked Intelligence" (Tapscott 1995, p. 6). A few years later, *e-business* was a guiding concept for business models with networked IT applications. In this context, concepts were developed in various fields of application such as *e-commerce* for relationships with customers, *e-procurement* for relationships with business partners and *e-government* for relationships between public authorities among themselves and with their stakeholders.

Many of today's innovations in the field of networked IT originate in the e-business era. However, about twenty years after the creation of the term *e-business,* the range of possibilities and the concepts and instruments of information technology have expanded in such a way that several experts proclaimed a new epoch almost simultaneously: Brynjolfsson and McAfee (2014) speak of the Second Machine Age and Schwab (2016) of the Fourth Industrial Revolution. The drivers include mobile internet usage, where IT-based functionality is ubiquitous and no longer tied to the use of PCs or laptops. Other fundamental technological developments are *artificial intelligence* and *blockchain*. The qualitative leaps of innovation that result from the increasing integration of IT into their environment also have a particular significance. This integration is reflected in the term *Internet of Things* (IoT): The Internet of Things is the integration of objects enhanced by information technology into the Internet. These items are often referred to as *smart*, such as smartcard, smartwatch, or smart home.

IT that is interlinked and integrated in its environment is an integral part or an indispensable prerequisite in many other fields of innovation: in robotics, in autonomous vehicles and planes, 3D printing, nanotechnology, biotechnology, renewable energies and more. Schwab stated that "the fundamental difference between the Fourth Industrial Revolution and the earlier revolutions is the close connection of these technologies and their interactions across the physical, digital and biological sphere" (2016, p. 18–19).

The recent expansions of IT and its new interactions with the environment have since the 2010s often been combined under the popular term digitalization. In this context, digitalization can be clearly understood as a collective term for economic and societal developments, which are very much determined by new technology applications and concepts. These are based on IT solutions that are ubiquitous and can also act independently, as a result of mobile technologies, cloud computing and digital representation of people, organizations and things. Digitalization extends personal and organizational forms of action in a variety of ways, e.g. in personal interaction (social media), in the design of business models (e-business), in the understanding of ownership (sharing economy) or in the creation of trust (blockchain).

From Schwab's understanding of the far-reaching changes caused by the *Fourth Industrial Revolution*, the following broad understanding of the popular concept of digital transformation can be deduced: *Digital transformation* refers to the change in different areas of life through increasing digitalization, also in connection with other technologies. The high speed and parallelism of several developments create dynamics and complexity that render reliable forecasting and planning difficult. In terms of the economy as a whole, the change in existing economic structures is intended, in case of a single company, the transformation of its own business model. The drivers for digital transformation in the economy are changing customer expectations, digitally revalued products, innovative forms of cooperation and new operating models.

The aspect of the *digital*, already used in the year 1995 by Don Tapscott in the term *digital economy*, is largely congruent with the meanings that underlie the notions of "e-" that have been popular since the end of the 1990s, e.g. e-business. In practice, these terms are used with different meanings and cannot be sharply differentiated. Some players understand terms with "e-" primarily in the context of Internet applications, others have a technology-neutral understanding. The Swiss State Secretariat for Economy Seco (undated) uses the terms synonymously, according to which *digital economy* and *e-economy* are synonymous.

The terms *digital business* and *e-business* can also be regarded as synonyms. However, there are two meanings, which should be distinguished: in a *company-internal* perspective, *digital business*—or *e-business*—supports the relationships and processes of a company with its business partners, customers and employees through networked information technology. In contrast, in a *cross-company* perspective, *digital business* refers to industries, companies or business models that provide their services in digital form, or where interlinked information technology is a defining feature of value creation. Examples are search engines, online marketplaces or media streaming.

3 Mobile Devices—Drivers of the Digital Transformation

Since Apple launched the iPhone in the year 2007, the use of the Internet via mobile devices has steadily increased. Globally speaking, in November 2016, the smartphone surpassed the mostly stationary devices such as the desktop and laptop for Internet access (StatCounter 2017). The ever-increasing distribution of smartphones and their ubiquitous and easy mobile use make them a key driver of digital transformation. With mobile wireless networks and simple functional expansion through applications (apps), the potential for use in innovative business models is virtually unlimited. In the following, we refer to the three factors *wireless networks*, *mobile devices* and *functional expansion* and give examples of how the interaction of the factors can be used in business models.

The first factor is *wireless networks*. These include the mobile networks of telecommunications companies and the local networks of companies and households. The cost of their use has been steadily decreasing over the last ten years. Prior to the introduction of the iPhone, the transfer of one megabyte of data via the UMTS network of Swisscom cost up to 10 Swiss francs. Data volumes were not a common component of a mobile phone subscription. In the year 2017, the subscriptions usually contain data volumes of several gigabytes or even data flat rates. In the years in between, the mobile networks became faster by a factor of a thousand: from around 300 Kbit/s with UMTS to 300 Mbit/s with LTE (Fettweis and Alamouti 2014). There has been a similar development with regard to the local wireless networks of companies and households that are mostly based on wireless LAN technologies. These have also become increasingly faster in the last ten years and they can be used cost-effectively or even free of charge.

The second factor is the *mobile devices themselves*. The ARM processors that are built into a smartphone today can host the transistors of an Intel processor in a desktop of ten years ago several times over (Wikipedia 2017). The performance of today's smartphones allows the execution of functions that were performed a few years ago on stationary computers: for example, surfing the Internet or playing high-resolution videos. Moreover, smartphones have lower power consumption and often weigh less than 150 g—characteristics that are decisive for the mobility of the devices. Unlike a laptop, a smartphone is almost continuously in operation and connected to a network. Regardless of time and place, it is always at hand and immediately ready for the input and output of data. In addition, mobile devices have sensors and other functions, which a stationary terminal usually does not have, e.g. a GPS (Global Positioning System) receiver for the determination of the exact location, an acceleration sensor for capturing movement in a room or a camera for recording pictures and videos. However, the handiness of mobile devices has its disadvantages: Because of their small size and data input via touchscreen, they cannot be used as extensively as stationary equipment with large screens, mouse and keyboard.

These features make mobile devices attractive for business models, when it is a matter of digitally *recording or issuing* data and information, automated as much as possible, at any time and directly with the user. *Mobile devices can automate business processes* that up to now have been heavily influenced by manual or paper-based activities. An example is the model of FAIRTIQ (2017), which will be explained below.

The third factor is the *simple function extension* of mobile devices with applications (apps). Millions of apps are available for the mobile operating systems Apple iOS and Google Android (Cuadrado and Dueñas 2012). The apps can be offered by anyone and provide features that permit the use of a smartphone or tablet for similar purposes as a stationary terminal, e.g. for e-banking or online shopping, and also for functions that are set up based on the specific features described above, such as sensors or the continuous operation of smartphones.

Below, selected examples of business models are described which are based on the capabilities of the three factors mentioned, and which can only be implemented and fully digitalized by the combination of these factors.

In some regions of Switzerland, passengers can use public transport with the FAIRTIQ app without first having to buy a ticket (FAIRTIQ 2017). The passenger must first install the app and then register via FAIRTIQ. She or he can then enter any means of transport and "check in" via the app. The route is automatically recorded using GPS data. At the end of the journey, the passenger has to "check out" again. If he or she forgets this, the app has a special function: It continually checks whether the passenger is still travelling on a route. For this purpose, the network of the FAIRTIQ

Control Center compares the data of all means of transport in the vicinity with the data from the smartphone. If the data does not correspond to a certain degree, the app will prompt the customer to confirm the end of his or her journey. Immediately after checkout, the cheapest fare is automatically charged for the route traveled.

The business model of Uber uses data which is similar to FAIRTIQ. With the help of the GPS data from the passenger's smartphone, the route is determined and settled. In the past, Uber used this data not only for the actual business purpose, but also for the unmasking of investigators who wanted to detect illegal drivers. This was known as "Greyballing" (Isaac 2017): In many cities, the business model of Uber is politically and legally controversial, as the drivers are often neither specially trained nor officially licensed, unlike taxi drivers. In some cities, the prosecution authorities searched specifically for drivers by having their investigators use the driving services Uber. In order to counteract this approach, Uber analyzed the collected data and presented the passengers incorrect data on available drivers via the app, if the passenger had lingered at length near police stations. The assumption was that these people might have been potential investigators.

Smartphones also have their use for customer loyalty programs. In its customer app, the CSS Insurance uses the acceleration sensor to count steps (CSS Krankenversicherung 2017). The more steps are recorded, the higher the reduction in the health insurance premium, because the more a person moves, the healthier he or she remains. The CSS aims to motivate the insured persons to participate in health prophylaxis and at the same time strengthen customer loyalty. The latter results from the premium that the customer loses when he or she changes the insurance provider.

4 E-Commerce: Fragmentation of the Supply Side Through Digital Transformation

Simply looking at e-commerce as a further distribution channel for products and services besides the existing ones has turned out not to be sufficient. In many industries, digital transformation is reflected in changes in the traditional distribution chains as a whole as well as in changes in the business models of the individual players. This chapter explains this development and shows that further momentum will develop in the years to come.

The starting point of the changes is the massive increase in transparency due to the World Wide Web. On the one hand, it has allowed customers to get a much clearer picture of the available offer in the preparation of purchases, which has helped them to shop more competently. On the other hand, the process of informative purchase preparation has been decoupled from the actual purchase at a finally selected provider. The collection of target-group-specific assortments as well as information and advisory services—classic trading functions—are now provided by numerous players online in different forms: search engines, price comparison services, online marketplaces, manufacturers, retailers, topic pages and social media. Never before have so many sources of information been offered to potential buyers. The provider side has lost its information monopoly: Along with this, customers and other experts now also communicate their assessments to others via social media and influence purchasing decisions.

The information availability, decoupled from the traditional dealer, has caused the *fragmentation of the customer journey*. The simple ROPO schema—*research online, purchase offline*—has been enhanced by customer behavior, in which some purchase transactions are resolved into numerous individual steps with advance and backward jumps to various sources of information. The process can drag on for many weeks, for example, before a vacation trip. The single contact in such a fragmented customer journey is called the *touch point*. Suppliers have to accept that they can no longer operate the purchase process as a whole, but are only one of several touch points.

Touch points have become a pivotal point in sales. The challenge is twofold: On the one hand, it involves introducing interested parties to the touch point, i.e. to find access to potential customers. On the Internet, the challenge is to generate enough traffic for one's own website and in stationary trading there is the need for sufficient customer frequency in the retail stores. On the other hand, it is necessary to design one's own touch point in such a way that the prospective buyer does not move on, but rather takes a purchase decision. It is neither necessary that the purchase is carried out in the same channel in which the purchase decision was effected, nor that the purchase takes place where the decision was made. An example of the first case is a customer who informs him-/herself on her/his smartphone about an offer, then checks its availability in a particular shop, and then purposefully seeks out this particular shop to complete the purchase. In such a *cross-channel scenario*, the customer uses two channels from the same provider in a purchase process. An example of the second case is a customer who informs him-/herself on a price comparison platform about the available offers, decides on a provider, is passed on through a link to this specific online shop and carries out the order. In this second scenario, the intermediary price comparison platform receives compensation from the provider for the fact that it has forwarded a prospective buyer with a high likelihood of purchase.

These examples show that digital transformation not only leads to a fragmentation of the customer's purchasing process, but also that the services on the provider side are unbundled. Although, there are still purchasing situations in which the complete purchase process is performed with a single provider, e.g. buying a book in a bookstore or the habitual buying of breakfast rolls at the bakery around the corner. A large part of the innovations in commerce and especially in e-commerce, however, is precisely that providers focus on individual trading functions and that the other functions are provided by third parties. This focus has created new business models that have a previously unknown level of performance. For example, in sales and marketing for the travel industry, the following business models have emerged[2]:

- Horizontal search engines (e.g. Google)
- Vertical search engines (e.g. TripAdvisor or Trivago)

---

[2] A detailed description of digital transformation in the accommodation industry can be found in the e-Commerce-Report Schweiz 2017 (Wölfle and Leimstoll 2017).

- Theme or destination platforms (e.g. http://www.myswitzerland.com)
- Online booking platforms (e.g. http://www.booking.com)
- Travel events based on dynamic packaging (e.g. http://www.hotelplan.ch)
- Flash sales providers (e.g. http://www.deindeal.ch)
- Sharing platforms (e.g. Airbnb).

At first glance, it is not easy to see the differences between the various business models. They are each highly relevant for a particular aspect of a purchasing process and the consistent focus on this differentiating aspect is the basis of their high attractiveness for a market segment. A booking platform for accommodation, for example, offers a wide range of available offers with comprehensive information, including customer evaluations and special offers. Once a traveler's data are collected, he or she can easily book the accommodation with a few mouse clicks. The offer of a booking platform is limited, however, by the fact that with each provider the conditions for the direct booking facility must first be worked out. In this point, vertical search engines, also called meta-search engines, differ: They can present a wider offer to the interested parties, also competing offers from several intermediaries for the same hotel. This provides a benefit for customers who appreciate a very wide selection or are looking for the cheapest price. However, meta-search engines link to another website for the booking process. Guests usually then have to familiarize themselves with this website, and have to enter their data.

The popular Internet business models are intermediaries in sales, which deliver the assortment function as well as stimulation, information and consulting services. These services are beyond the scope of the service of the dealers or the providers themselves. Other functions from the classical functional package of trade, however, are also provided by specialists, as predicted by Albers and Peters (1997). Examples are logistics functions for physical products, payment processing and the receivable risk or the provision of warranty and repair services. It is no longer a complete sales concept, but rather, each function has become the subject of competition, optimization and innovation. Innovative sales concepts emerge from new combinations of functions, each of which can represent an innovation (Leimstoll and Wölfle 2014).

The competition, which is associated with specialization, means that individual companies are less able to keep pace with only the help of their internal resources. In many market segments, performance leadership is currently achieved by so-called *platforms* that aggregate services from multiple providers. There are various forms of manifestations: In addition to search engines and comparison platforms, there are also hybrid providers. A brand-defining online hybrid provider integrates offers from selected third party providers into his shop. These open two-sided online marketplaces bring together providers and consumers to platforms such as Zalando. They also involve other players from the fashion industry, such as logistics service providers, and market influencers like bloggers. The benefits of the platforms in an increasingly diverse and complex world are the creation of transparency, unifying offers, setting standards for routine operations and coordinating many stakeholders in an efficient manner (Wölfle and Leimstoll 2017, p. 23). Successful platforms achieve high, positive economies of scale and can thus take on large investments in their infrastructure and performance.

In view of the ongoing innovations, a high level of dynamism has to be expected in the coming years. As soon as they have mastered the challenges of mobile business, the providers will have to face new paradigm shifts: interactions via chat and voice, the latter neither with a keyboard nor with a screen. In a few years, digital assistants on smartphones or on devices such as Amazon Echo could become the hub of everyday operations and related purchases. Similar to operating systems for smartphones, it is to be expected that in the case of digital assistants only a few platforms will be able to prevail worldwide. These would then become a further intermediary between providers and customers, imposing the rules of the platform on all involved and gaining an advantage from their market transparency.

Suppliers will be increasingly unable to escape this development. The challenge for them is to find a suitable combination of self-provided and third party performed value-added functions. The business model must permit demand to be triggered or found, and on the other hand, it must achieve a viable cost-benefit ratio.

5 E-Procurement: Digital Business-to-Business Networks Support Transformation in Purchasing

Digital business-to-business networks (B2B networks) are currently the pre-dominant solution scenario when it comes to electronically supporting or even integrating business processes between procurement organizations and their suppliers. Platforms offered by service providers such as SAP Ariba, IBM Sterling Collaboration Network, Perfect Commerce or Tradeshift connect a variety of procurement organizations with their suppliers. They simplify the electronic exchange of data between the business partners and thus expand the core functionality of the established Enterprise Resource Planning systems (ERP systems) with digital B2B processes. This makes it possible to digitally transform business relationships and company-wide processes in a targeted and efficient manner. B2B networks clearly help to make purchasing more effective and efficient and to increase the value contribution of the procurement function for the company.

Once companies were able to support their internal operational processes with information technology, in transaction-intensive industries in the 1980s the need grew to digitize business processes across companies in order to eliminate the manual entry of business documents, such as purchase orders, delivery notices, order confirmations and invoices, and to speed up processing. Various efforts eventually led to the development of the EDI standards ANSI X12 and UN/EDIFACT (Liegl 2009). These consolidated the requirements of various industries and defined the contents and structures of business documents as well as the processes for their electronic exchange. Over time, however, it became clear that these standards were not implemented uniformly. The reasons for this were the different interpretations of the directives, the lack of experience of the implementation partners, and often simply the lack of time that would have been necessary to carry out detailed clarifications or to wait for amendments to the standards.

Around the turn of the millennium, additional XML-based standards were added, such as cXML, xCBL, UBL, and UN/CEFACT XML. This growing number of standards also complicated the coordination between the business partners. In business relationships with thousands of transactions, a complex bilateral orchestration of processes and data structures may still be tolerated. In the case of smaller transaction volumes, however, it takes a lot of convincing to integrate business partners directly, because the cost-benefit ratio is not right for the supplier. Many business software providers are also lagging behind the requirement of their customers to find a simple way to implement integrated cross-company business processes. In addition, not all business relationships are established for a long-term duration, and in such cases, above all, flexible, efficient integration procedures are required.

The digital business-to-business networks have accepted this challenge. They translate the different data formats and convert the differences in interpretation into the data structures and contents required by the receiver. This enables the business documents and the following processes (e.g. control of delivery date or invoice) to be processed automatically. B2B networks are striving to connect the largest possible number of suppliers and procurement organizations to each other. For the individual affiliated company, the perceived value of the network increases as current and future business partners increasingly handle their processes via the platform of the respective service provider. As a result, B2B networks enable an agile networking of companies and help to overcome the variety of standards and their heterogeneous implementation. The number of business partners who can be reached can be expanded if a B2B network connects its integration platform to other networks. This so-called interoperability ensures that data can be exchanged between networks without loss of information. Although interoperability between networks is still not common, the electronic exchange of data in B2B relationships as the first stage of digitalization has either already been realized by many companies or will be easy to achieve, due to the B2B networks.

In their role as service providers between suppliers and procurement organizations, the B2B networks now offer additional value-added functions and electronic processes that are usually not supported by traditional ERP systems (Fischer 2014). These are, for example, the processing of electronic requests for information (RFI) or for quote (RFQ), online auctions or the management of electronic product catalogues. These functions are offered as software-as-a-service in the cloud. It is also possible to evaluate the transactions handled via the B2B network, which improves transparency and control capabilities. For suppliers, this has the advantage that they do not have to operate dozens or even hundreds of customer-specific supplier portals and thus benefit from functional consolidation, as in the case of data exchange. The above-mentioned extended functionalities are primarily used by large companies operating globally. Medium-sized companies still hesitate and use the B2B networks mainly for electronic data exchange (Tanner and Leimstoll 2015).

The world's leading business software provider, SAP, recognized that the integration of business partners into B2B processes via a network model is more efficient and more promising than direct connection via the ERP functionality. Hence, SAP acquired the world's largest B2B network, Ariba, for US $4.2 billion in the year 2012 (SAP 2012). Through this B2B network, the interlinking of companies is much simpler and more standardized.

The cooperation in 2017 between IBM and SAP Ariba (IBM 2017) demonstrates the increasingly relevant topic of digitalization in purchasing. Based on structured and unstructured data, sourcing managers are to be supported by IBM Watson's technology in their search for supply sources and thus enabled to make better decisions. By means of "cognitive procurement" (IBM 2017), purchasing managers should become aware of delivery risks and get recommendations for the optimization of procurement by considering existing supplier and transaction data, combined with, for example weather data, geo-political information and raw material prices.

Although B2B networks are the prevalent solution scenario in the electronic collaboration between business partners, small and medium-sized companies that could benefit greatly from their services are reluctant to use them. The reasons for this are manifold. They range from inappropriate price models, from uncertainty regarding legal issues, from concerns about data protection and data security, to internal cultural and functional barriers or insufficient support from the business software systems used. The design field is demanding, especially since functionally, geographically and technically different levers have to be considered. This requires a competent conceptual approach for which the procurement departments often lack the necessary resources and competences (Pesonen and Smolander 2011; Tanner and Leimstoll 2015). Further, the management's attention in digital transformation projects rarely focuses on procurement. However, we can assume that with the increased proliferation and use of B2B networks and with the growing pressure to be efficient due to global competition, the threshold for SMEs to connect with B2B networks will be reduced.

6 E-Organization: Flexible Business Software as the Backbone of Corporate Transformation

The importance of integrated business software systems for the digital transformation of companies and other organizations is often underestimated. This might have something to do with the fact that such systems have been in use for decades, and are therefore omnipresent in the workplace and are a part of the basic equipment. Connecting them with innovative processes of change in the company does not seem to be obvious, especially as the dinosaurs of electronic support for business processes seem rather to cement structures instead of making them more flexible. Both can be the case, but the developments of recent years clearly speak for the fact that modern business software systems have become more flexible in many respects. They

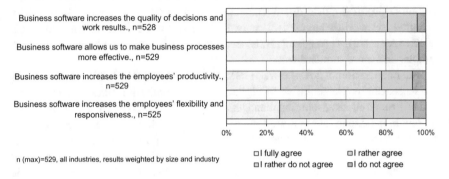

Fig. 1 Assessment of the usefulness of business software

may not exactly be a driver of digital transformation, but they can be a hub for the innovative design and support of work-related value-added processes.

Business software systems is a very broad term. This category includes process-specific business applications, Enterprise Resource Planning systems (ERP systems), and e-business systems to support cross-company processes (Kees 2015, p. 12 ff.). The following statements refer mainly to integrated ERP and e-business systems.

The importance of integrated business software systems within digital transformation is primarily the result of its position as a central data receiver and data supplier for the integration of value-added processes across companies, for example in the context of Industry 4.0 or in the service sector within the smart services. Thus, they provide a central point of connection and consolidation for distributed business logic and data management. Studies prove that mature systems with their database contribute to the productivity and flexibility of business processes. An investigation into small and medium-sized enterprises (SMEs) in Switzerland shows that in about 80% of companies positive benefits are achieved with the use of business software, such as the improvement of decision quality, more effective design of business processes or an increase in productivity or flexibility (see Fig. 1). The study also refers to the quality of the data: almost all of the nearly 550 SMEs surveyed say that their data can always be kept up to date with the help of their business software; around 95% agree that their software provides the necessary data (Leimstoll and Quade 2016).

Both the processes supported by business software and the ERP systems themselves become more flexible (Gronau 2014, p. 42 ff.): They offer more adaptability to specific business processes while keeping the maintainability, up to the profitable development of individual software due to matured frameworks and development environments. Individually programmed functional additions do not necessarily originate from the actual provider of the business software. In particular, traditional ERP providers have always been rather reserved with regard to data analysis and the rapid delivery of real-time key figures. Historically, the separation of operational and analytical systems goes back to the specific requirements of analytical databases that the operational databases were not able to fulfill, neither technically nor conceptually. With the transition to in-memory computing (loading the data into large RAM memory) this division loses importance and data analysis can be carried out with operational data, due to significantly higher access speeds (Plattner and Zeier 2012). Direct access to physical database systems, for example via database interfaces such as Java Database Connectivity (JDBC), can also be very fast because they bypass the complex ERP logic.

A further very important aspect of flexibility, not only of business software, is the flexibility of the functional distribution. With the ongoing predominant client-server principle, functions and data management can be distributed in different ways on the server and client, depending on the requirements of the situation and the end-user device (Bahssas et al. 2015). For example, for terminal servers, the clients of the business software are operated as virtual clients on the server rather than on the end-user device. Similar, but somewhat more modern, is desktop virtualization. Here the server provides complete applications for use so that they do not have to be installed on the end-user device. In both cases, a standard application with input and output functionality (e.g. Citrix Client) is sufficient for the use of the device. Such architectures relieve the processing capacity of the end-user devices, reduce the effort of their installation and maintenance, permit the rapid connection of additional users and, last but not least, increase data protection and data security, because no data is stored on the end-user device.

Web-based virtualizations work in a similar way. They also pursue the goal of running the complete business logic on the server, because the universal clients used (typically browsers, if necessary, enhanced by Rich Internet Applications, or Java-based clients) do not have any application-specific logic.

Common to all types of application virtualization is that they make software and hardware more independent of each other. This creates freedom in the selection of end-user devices and operating systems. The business software, or at least parts of it, can then be used with mobile devices, for example, or on platforms (e.g. Apple computer) for which they were not originally intended. All the virtualization concepts also have in common the fact that they require access to a network and, of course, accessibility to the servers. As a result, it is proving advantageous that more and more business software products are being offered as a cloud solution and thus are intended for use via the Internet.

If business software systems or parts thereof are used offline as well, then functionality must be available on the end-user device. This is the case for mobile apps. Another form of selective use of individual ERP functions or rather the selective interaction with ERP systems are applications in the area of the Internet of Things, e.g. smart products that order the required consumables themselves. A physical, networked device triggers the order automatically. The Amazon Dash Replenishment service is a well-known example.

7 E-Organization: The Merging of Collaboration and Process Management

Digital change is also increasingly transforming process management. This is shown in part by the fact that processes of collaboration in companies are increasingly being digitalized in the form of workflows.

Successful collaboration in companies is an important factor in order to be innovative and able to solve problems efficiently (Paulus and Nijstad 2003, p. 110). The collaboration always follows a sequence of activities, as an implicit or explicit process (Weske 2012, p. 4). Implicit, in this context, means that the participants have a process flow in their heads, which does not have to be identical for all parties involved. Collaboration then takes place according to these personal ideas. Explicit means that there is a formal process description in the form of process models, implementation documents, checklists or similar, where all parties orient themselves.

The digitalization of processes in the field of collaboration is an important measure to ensure effectiveness and efficiency in the collaboration within a company and to be able to act flexibly (Dorn et al. 2012, p. 73). Of particular importance is the (partial) automation of the exchange of information (Nof et al. 2015, pp. 13–20). As a result, process management becomes an integral part of the collaboration.

Collaboration in companies can take many forms. This may involve *individual situations*, on the one hand, such as the development of a solution to a problem that occurs for the first time, or the development of new concepts. On the other hand, collaboration can consist of *routine processes* in which the same activities have to be processed repeatedly, e.g. when a new employee joins a company and predefined work needs to be done in various areas or departments (on-boarding). In this case, the process is usually formalized in the form of documentation.

The proportion of documented processes in companies is steadily increasing. Process Management Systems model and manage them and make them available to employees (Weske 2012, pp. 305–333). The maturity of the documented processes, however, differs greatly: It ranges from simple checklists to modeled and guided processes. Several departments may be involved, for example in an on-boarding process. Very often, checklists are used to ensure that all necessary tasks are processed. The cooperation then takes place through the handover of the checklist from one department to the next. Tasks such as the provision of IT resources or access cards are usually only controlled by checklists. There is no central overview of the status, lead times, etc., of a process.

A process can also be regulated by e-mail, whereby parts can be parallelized. The relevant documents are then sent by e-mail and the points to be dealt with are "ticked off" in the documents. This results in almost the same shortcomings concerning the transparency of the process as with the printed checklists.

Such practices are common, although many companies today already use collaboration platforms such as Microsoft SharePoint or similar ones to share documents or work together on content. Such platforms also contain tools that allow the automation of processes, either wholly or partially (so-called workflow engines). Meanwhile,

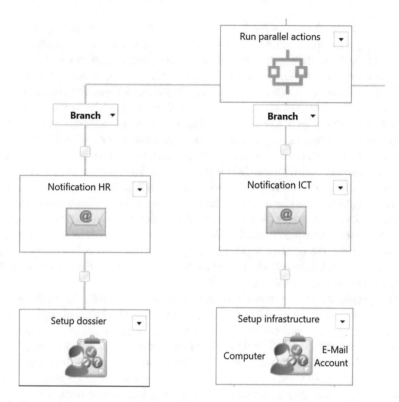

Fig. 2 Excerpt from an on-boarding process with Nintex in MS SharePoint

From: noreply@fhnw.ch [mailto:noreply@fhnw.ch]
Sent: Montag, 19. Juni 2017 07:59
To: HR <hr@fhnw.ch>
Subject: Mutation of Johnson Claire

A task is assigned to you:

Mutation: Johnson Claire

Click here to work on the task.

Fig. 3 Excerpt from a task by e-mail from the on-boarding process with Nintex in MS SharePoint

such tools have a high maturity level, which also allows non-experts to create work-flows. With minimal effort, processes can be programmed graphically with just a few clicks. Figure 2 shows how an on-boarding process can be implemented very quickly as a workflow. The tasks are parallelized by sending and controlling subtasks automatically via e-mail (see Fig. 3).

The controlled workflow ensures that all tasks are processed and escalation levels defined, for example, that the task "setup dossier" must be completed within two days. If this is not the case, a reminder will be sent automatically and after two more days, the higher responsible person will be informed. The state of the workflow can be viewed at any time. After completion of the process, evaluations of the effectiveness are possible and the retrieval of lead times permits efficiency to be controlled.

Thus, the use of workflows brings significant advantages for companies and employees. First, the processes are *documented* by the modeling of the workflow, whereby the knowledge of a process is not lost even if a knowledge carrier is lost. Secondly, the *effectiveness and efficiency* of the collaboration are ensured: In the sense of professional process management, it is guaranteed, at all times, that employees know the currently valid process, abide by the procedure and thus carry out the collaboration in the correct manner (Cichocki et al. 2012, p. 20). They no longer need to study checklists, flowcharts, and the like to be sure that they follow the guidelines. Third, workflow documentation and control together create a high *process transparency*. Fourth, there are *no shadow documents* (e.g. copies of the original process documents from the process management systems) that are no longer up-to-date. The explanations for the activities are embedded in the tasks of the workflow, so that further documents are usually no longer necessary.

The specific feature of automating processes through workflows is that the process management system is not visible to the people involved in the collaboration. Thus, process management and collaboration merge and employees can concentrate on the actual content of the collaboration. For this reason, only a little to no training is necessary. Because of the simplicity of modeling such processes, the creation of a workflow can also be worthwhile for activities that do not take place very often. The creation often takes only a few minutes, whereas the otherwise necessary merging of results and the manual administration of the collaboration often take much more time and are prone to potential errors.

8 E-Organization: Digital Transformation Needs Hybrid Project Methods

In order to ensure that companies do not lose their connection in the digital age, they must decide which new digital technologies they want to use to modernize their business models and with which project methods to develop them. An example of this is the development of digital distribution channels such as online shops and mobile apps, and their integration into the existing sales processes. Moreover, the shortest possible time-to-market is becoming a key success factor.

Given such high standards, traditional project methods with their relatively rigid project organizations, hierarchy structures and fixed phase organization, such as Royce's Waterfall Model, first published in 1970 (Royce 1987), reach their limits and require realignment. In this context, agile project methods such as Scrum (Schwaber 1997) or Design Thinking (Buchanan 1992) gain in importance. Unlike the classic waterfall model, they rely on frequent iterations as well as on an intensive exchange and proximity between the participants. They also aim to bring innovations to market as quickly as possible.

In *classically* managed projects, the work is typically carried out in phases (project initialization, project planning, project implementation, project completion). For the purpose of controlling the implementation, there are mainly hierarchically-structured roles such as the project manager, the project team, the steering committee, and the customers. These are complemented if necessary by further roles for professional support (quality assurance, test management). The project manager is responsible for project planning and controlling and thus for the success of the project.

The classically managed project plans have undisputedly many *advantages*, which has led to these models of procedures—with all their variations—being the usual way of carrying out projects today. They are relatively easy to manage, the planning and progress monitoring is carried out with established instruments and methods, and the model is easy to understand and communicate. However, a major disadvantage is the lack of flexibility in relation to late changes in the project progression, because these models assume that the requirements and basic conditions are known from the outset and do not change during the project implementation. If the situation is given from the outset, for example, in the procurement of standard software for financial accounting, classic project management works very well.

In *agilely* led projects, deadlines and costs are fixed at the beginning. The requirements are outlined vaguely in the sense of a solution space (scope), as they are often not clearly delineated or are even partially unknown at the beginning of the project. Scrum (Schwaber and Sutherland 2016) is one of the most important representatives of agile project management. At the center of the matter is the scrum team, which is responsible as a team for the success of the project. It consists of a development team, the scrum master and the product owner. These three roles and the people behind these are responsible for the development and delivery of the software product (product increment). The development team usually consists of several people (programmers, testers, designers) who take responsibility for their tasks. The scrum master provides suitable working conditions and ensures that the rules and principles of the scrum process are adhered to. The product owner is responsible for the economic success of the product. She or he represents the stakeholders, in other words, the persons affected and involved in a project. The product owner also coordinates all requirements for the product.

At the start of a project, the product owner determines all the customer and stakeholder requirements known at the time, and records them in a list (product backlog). The most important items are at the top of the list. The prioritization is not primarily based on the costs and deadlines, but focuses on maximizing the value that is to be created for the company or the stakeholders in the fields of expertise. The items are divided into blocks by their order of precedence and are gradually processed by the team in stages, the so-called sprints. Sprints usually last thirty days. Within a sprint, the team members decide autonomously how they want to proceed and meet every day to exchange experiences (daily scrum). At the end of a sprint, the development team presents the new functions (product increment) in a usable form to the product owner. Once the sprint is completed, the next stage starts with the re-prioritization of the items from the product backlog and the subsequent sprint planning.

Agile project management can show its strengths if there are many unknown premises during the planning phase or premises that still need to be worked out in the project. Examples are projects for process improvement or software projects such as the redesign of an online customer portal. An iterative approach using sprints makes for more transparency and entails regular product deliveries (product increment). Possible obstacles become visible sooner and, if required, the product can be changed more quickly by an incremental procedure. In the future, agile methods are likely to be increasingly used in projects with a strong customer relationship and which have the demand for a short time-to-market. The agile approach, however, also requires much higher control and administration efforts.

Thus, both methods, classic and agile, have their strengths and weaknesses, so that neither the classic nor the agile procedure model is preferable to the other. Rather, it is a matter of finding the right mix of both worlds, and skillfully combining the different approaches. The experts often use the term "hybrid approach model" (Kirchhof and Kraft 2012) for this purpose. Further arguments for the *hybrid* approach result from the fact that agile approach models cannot be implemented companywide from one day to the next. Usually, therefore, the process only starts in a sub-area, typically in software development, to gain initial experience. If agile methods are used in one area, and classical methods in another, projects that require interdisciplinary collaboration across different departments must coordinate the work and results from the different approaches.

In *hybrid* projects the approach is similar to the purely agile project approach, as the flexible handling of changes, the focus on market requirements as well as the matching of results to the target group are all supported. The classical project leader continues to be responsible for the co-ordination of the project and for the achievement of the defined target within the given time and in line with the budget. The project manager is accountable to the company's management for the project process, and responsible for the overall project. From the agile approach, particularly the continuous exchange of information between the participants during the course of the project is adopted. During the project planning, the project manager considers which task packages are suitable to be worked on in an agilely conducted sub-project. To ensure the exchange of information, the scrum masters and the product owners will participate in the meetings of the traditional project team and allow feedback from other areas to be incorporated into their sprint planning. With open communication, decisions become more transparent, the understanding of other business areas and projects rises and acceptance of the joint approach is increased.

9   Conclusion

The objective of this contribution is to clarify the role of e-business in the era of digital transformation. After a brief definition of terms, the current developments in the classical e-business perspectives of e-commerce, e-procurement and e-organization are shown. The analysis makes it clear that taking different perspectives of e-business is not only still helpful in order to identify and stimulate meaningful processes of change in and around a company, but it also allows further reaching transformations in the economy to be analyzed and explained, especially the division of labour along the value chain or in value-added networks.

Digital transformation, as it is being discussed today, is characterized by far-reaching effects on the economy and society, a significantly increased rate of change, and an even greater convergence or merging of information technology with many other areas of technology. In the areas of e-commerce and e-procurement, digital transformation is reflected in an intensified cooperation between companies. In e-commerce, this results from the unbundling of the classic merchant functions and their distribution to specialized companies. In e-procurement, digital B2B networks facilitate the entry into the electronic exchange of documents. In both areas, more and more methods of analyzing comprehensive data are used.

The area of e-organization is dominated by the topic of agility, that is to say, the ability to adapt quickly to changes. On the one hand, this is necessary for companies to make digital change controllable. Therefore, a transition to agile or hybrid approach models can be useful for project management within the company. On the other hand, developments in the field of business software are providing solutions that are helping companies become more agile, such as applications from the cloud. The workflow management area is also characterized by more agile applications that can be used to design workflows very quickly.

This results in a variety of starting points for future research and development projects. Due to the dynamics resulting from the current development at different levels and in a wider range of disciplines, the need arises to observe and reflect the application possibilities of new technologies and to transfer them to other fields of application. It is to be expected that this will require an increasingly interdisciplinary research approach.

References

Albers S, Peters K (1997) Die Wertschöpfungskette des Handels im Zeitalter des Electronic Commerce, in Marketing ZFP, 19. Jahrgang 1997, Nr. 2, S 69–80

Bahssas DM, AlBar AM, Hoque R (2015) Enterprise resource planning (ERP) systems: design, trends and deployment. Int Technol Manage Rev 5:72–81

Brynjolfsson E, McAfee A (2014) The second machine age: wie die nächste digitale Revolution unser aller Leben verändern wird. Plassen/Börsenmedien AG, Kulmbach

Buchanan R (1992) Wicked problems in design thinking. Des Issues 8(2):5–21

Cichocki A, Ansari HA, Rusinkiewicz M, Woelk D (2012) Workflow and process automation: concepts and technology. Springer Science & Business Media

CSS Krankenversicherung (2017) myStep. In: myStep. https://www.css.ch/de/home/privatpersonen/kontakt_service/mycss/mystep.html. Accessed 10 Apr 2017

Cuadrado F, Dueñas JC (2012) Mobile application stores: success factors, existing approaches, and future developments. IEEE Commun Mag 50:160–167. https://doi.org/10.1109/MCOM.2012. 6353696

Dorn C, Taylor RN, Dustdar S (2012) Flexible social workflows: collaborations as human architecture. IEEE Internet Comput 16(2):72–77. https://doi.org/10.1109/MIC.2012.33

FAIRTIQ (2017) Entdecken Sie die einfachste Fahrkarte der Schweiz.| FAIRTIQ. In: FAIRTIQ. https://fairtiq.ch/de. Accessed 10 Apr 2017

Fettweis G, Alamouti S (2014) 5G: Personal mobile internet beyond what cellular did to telephony. IEEE Commun Mag 52:140–145. https://doi.org/10.1109/MCOM.2014.6736754

Fischer S (2014) Emerging Business Networks and the Future of Business Software. In: Brunetti G, Feld T, Heuser L, Schnitter J, Webel C (eds) Future business software: current trends in business software development. Springer International Publishing, Cham, pp 3–14

Gronau N (2014) Enterprise resource planning: architektur, funktionen und management von ERP-systemen, 3rd edn. Oldenbourg Wissenschaftsverlag, München

IBM (2017) SAP Ariba and IBM join forces to transform procurement with SAP Leonardo and IBM Watson. http://www-03.ibm.com/press/us/en/pressrelease/52405.wss. Accessed 19 May 2017

Isaac M (2017) How Uber Deceives the Authorities Worldwide. N. Y. Times

Kees A (2015) Open source enterprise software. Springer Fachmedien Wiesbaden, Wiesbaden

Kirchhof M, Kraft B (2012) Hybrides Vorgehensmodell: Agile und klassische Methoden im Projekt passend kombinieren. https://www.projektmagazin.de/artikel/agile-und-klassische-methoden-im-projekt-passend-kombinieren_1069867#cut-off. Accessed 10 Apr 2017

Leimstoll U, Quade M (2016) Special Business-Software-Studie 2016: Business-Software—ein Erfolgsfaktor in Schweizer KMUs? Netzwoche 8, S 1–8 (Special)

Leimstoll U, Wölfle R (2014) Auswirkungen des E-Commerce auf die Wertschöpfungsstrukturen im B2C-Einzelhandel: Erkenntnisse aus der Schweiz. Multikonferenz Wirtschaftsinformatik 2014 (MKWI 2014), Paderborn

Liegl P (2009) Business documents for inter-organizational business processes. PhD thesis submitted at the vienna university of technology faculty of informatics

Nof SY, Ceroni J, Jeong W, Moghaddam M (2015) Revolutionizing collaboration through e-Work, e-Business, and e-Service. Springer

Paulus PB, Nijstad BA (2003) Group creativity: innovation through collaboration. Oxford University Press

Pesonen T, Smolander K (2011) Observations on e-Business implementation capabilities in heterogeneous business networks. In: Skersys T et al. (ed) I3E 2011, IFIP AICT 353, pp 212–226

Plattner H, Zeier A (2012) Desirability, feasibility, viability: the impact of in-memory. In-memory data management. Springer, Heidelberg, pp 3–19

Royce WW (1987) Managing the development of large software systems: concepts and techniques. In: Proceedings of the 9th international conference on software engineering. IEEE Computer Society Press

SAP (2012) SAP completes acquisition of Ariba, Inc. http://news.sap.com/sap-completes-acquisition-of-ariba-inc/. Accessed 15 Apr 2017

Schwab K (2016) Die vierte industrielle Revolution. Pantheon, München

Schwaber K, Sutherland J (2016) The scrum guide-the definitive guide to scrum: The rules of the game, July 2011. http://www.scrum.org/storage/scrumguides/ScrumGuide

Schwaber K (1997) Scrum development process. Business object design and implementation. Springer, London, S 117–134

Staatssekretariat für Wirtschaft Seco (ohne Datum) E-Economy. Abgerufen von. https://www. seco.admin.ch/seco/de/home/Standortfoerderung/KMU-Politik/E-Economy_E-Government/E-Economy.html

StatCounter (2017) Desktop vs mobile vs tablet market share Worldwide. In: StatCounter Glob. Stats. http://gs.statcounter.com/platform-market-share/desktop-mobile-tablet. Accessed 31 Mar 2017

Tanner C, Leimstoll U (2015) Einsatz von IT im Einkauf in Grossunternehmen in der Schweiz (Use of IT in procurement in large companies in Switzerland). Hochschule für Wirtschaft, Fachhochschule Nordwestschweiz, Basel

Tapscott D (1995) The digital economy—promise and peril in the age of networked intelligence. McGraw-Hill, New York

Weske M (2012) Business process management: concepts, languages, architectures, 2nd edn. Springer, Heidelberg, New York

Wikipedia (2017) Transistor count. Wikipedia

Wölfle R (2016) Digitale Transformation—eine begriffliche Standortbestimmung im Jahr 2016. Arbeitsberichte der Hochschule für Wirtschaft FHNW Nr. 100, Hochschule für Wirtschaft, Fachhochschule Nordwestschweiz, Basel

Wölfle R, Leimstoll U (2017) E-Commerce-Report Schweiz 2017: Digitalisierung im Vertrieb an Konsumenten, Eine qualitative Studie aus Sicht der Anbieter. Institut für Wirtschaftsinformatik, Hochschule für Wirtschaft, Fachhochschule Nordwestschweiz, Basel

Digitalizing B2B Business Processes—The Learnings from E-Invoicing

Christian Tanner and Sarah-Louise Richter

Abstract Digitalizing an existing business process often proves to be more complicated than expected. This article provides insights into obstacles and success factors when digitalizing a business process, using the example of the transition from a paper-based process of handling invoices to electronic invoicing. Since e-invoicing has gained significant momentum in recent years from a business perspective as well as from governments all over the world, it provides an interesting area in which to investigate digitalization. Drawing from input collected in more than 10 years of research on the topic of e-invoicing, the authors illustrate why digitalization is still not easily achieved, despite the obvious advantages, and elaborate on the key prerequisites for success. The results emphasize the importance of understanding the needs of one's business partners and working closely with them when developing solutions. Furthermore, systematic project management and change management are important. However, as much as there is no "one size fits all" solution, there is also none that will last forever. Rather, as the business environment changes and technology matures, there will be a need to re-assess processes and solutions from time to time. Most importantly, the human factor of change cannot be underestimated. Besides standard change management practices, companies should seize the opportunity to develop their employees through the digitalization effort by engaging them in projects and decision-making processes. Acquiring project management skills, expert knowledge and experience in innovating business processes will serve as an invaluable asset in the long term.

Keywords Digitalization · Business process · E-Invoicing · B2B · Success factors

C. Tanner (✉) · S.-L. Richter
Institute for Information Systems, University of Applied Sciences and Arts Northwestern
Switzerland, Peter Merian-Strasse 86, 4002 Basel, Switzerland
e-mail: christian.tanner@fhnw.ch

S.-L. Richter
e-mail: sarahlouise.richter@fhnw.ch

© Springer International Publishing AG 2018
R. Dornberger (ed.), *Business Information Systems and Technology 4.0*,
Studies in Systems, Decision and Control 141,
https://doi.org/10.1007/978-3-319-74322-6_7

1   Introduction

Digitalization offers huge potential to streamline business-to-business (B2B) processes in terms of the manual amount of work, material, cost and time. However, many initiatives to digitalize established business processes suffer considerable delays or fail altogether, eventually not delivering the results promised in the business case. Hereinafter, we will illustrate the digitalization of a business process by looking into the example of the transition from a paper-based process of handling invoices to electronic invoicing and derive useful learnings that can serve as guidelines in general. Therefore, we examine obstacles and success factors to uncover the underlying challenges of process digitalization before providing tips on how to prevent and address issues.

In addition to classical literature research, this article relies on insights gained in more than 10 years of intensive research into the topic of e-invoicing in the Swiss Digital Invoice-Forum (swissDIGIN). Since 2004, swissDIGIN has facilitated e-invoicing in Switzerland by connecting companies, service providers, governmental authorities and researchers in a community. As the legal frameworks are different across the world, the article focuses on e-invoicing in Europe and Switzerland as an illustrative example.

Figure 1 shows a framework of change through digitalization, which will be used to visualize and structure the content of this article.

Companies are influenced by the general business environment, classified by the factors of the PESTEL analysis (political, economic, social, technological,

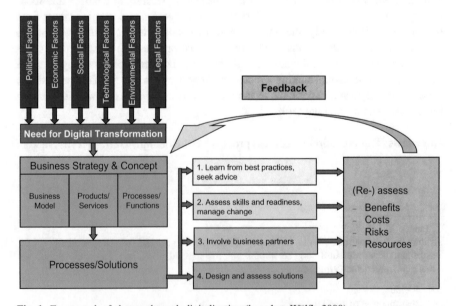

Fig. 1 Framework of change through digitalization (based on Wölfle 2000)

environmental and legal), which eventually trigger business transformation. The impact on strategy can range from disruptive changes such as new business models to moderate changes such as the optimization of processes. As a result, any change will call for new digital processes and solutions leading to a number of activities that, after assessment, will lead to implementation and continuous reassessment activities, affecting the business strategy and concept.

Based on the framework introduced in Fig. 1, the transition to electronic invoicing is, notably, not about disruptive change. Instead, it can be seen as a classic example for the digitalization of a business process that is managed in any given company or organization striving for more efficiency.

In the next section, we examine the maturation of e-invoicing over time, obstacles and key success factors. Thereafter, we provide an overview of recent developments before drawing a conclusion and pointing out the lessons learned.

2 The Need for Digital Transformation in Invoicing

In recent years, e-invoicing has gained significant momentum from a business perspective as well as from governments all over the world. Hence, not a single year in the last decade passed without being declared the year of the e-invoicing breakthrough (Ulrich 2002; Rombel 2007; Hawser 2009; Airey 2014). To understand today's obstacles as well as success factors that will be examined in the following sections, it is necessary to have a look at how e-invoicing developed. Following the political agenda and the economic potential to promote e-invoicing, the necessary framework conditions were established aiming at an efficient and transparent handling of invoices that contributes to a competitive economy (Tanner and Wölfle 2011).

We focus on the political, economic, technological and legal factors mentioned in the model introduced in Fig. 1. From a political perspective, e-invoicing has gained considerable attention for its economic benefits. Industry experts point out that by utilizing e-invoicing, the potential savings in the EU's public sector are estimated to be at least 40 billion Euro of which today, less than 10% is exploited (Koch 2017a). Together with the savings for the general economy, these figures amount to being relevant to the national economy. The European Union (EU) has set the goal to make e-invoicing the primary method of invoicing by 2020 (European Commission 2013). Therefore, the EU launched initiatives to drive the adoption of e-invoicing such as the Pan-European Public Procurement OnLine (PEPPOL) initiative seeking to enable interoperability between dissimilar systems through providing technical specifications being implemented into existing e-procurement applications (Pan-European Public Procurement OnLine 2017). Moreover, the United Nations Centre for Trade Facilitation and Electronic Business (UN/CEFACT), a body of the United Nations Economic Commission for Europe (UNECE), drives the harmonization of trade processes internationally.

From an economic perspective, implementing e-invoicing is attractive as it reduces "waste" (i.e. non-value-adding manual labor), offering significant potential for optimization for companies of any size (Kioses et al. 2007). Furthermore, e-invoicing processes are highly scalable, supporting growing businesses while, at the same time, reducing the need for manual work. Research findings suggest that companies, depending on the amount of paper invoices handled, could save up to 80% of their current costs related to the process (Koch 2017a). Contemporary cloud-based solutions are easy to implement, even for small and medium-size businesses (SMEs) with little need for investment or particular expert knowledge (Hayward 2013; Koch 2017b). Thus, if followed through in a straightforward fashion, payback periods of e-invoicing projects can be as short as 6 months (Mayer 2011; Koch 2017a). Further, findings show that e-invoicing significantly reduces the carbon footprint of an organization (Tenhunen and Penttinen 2010). Altogether, in globalized markets with fierce competition, e-invoicing can contribute to efficiency gains.

From the technological point of view, e-invoicing itself is not a recent invention, since companies trading in transaction-intensive industries such as retail and automotive started to send and receive invoices electronically through electronic data interchange (EDI) in the 1980s (Self 2008; Tanner and Wölfle 2011). However, the first EDI systems were rather complex and expensive to implement and maintain. Additionally, due to a lack of standardization, suppliers faced the situation of having to comply with their customers' preferences and deliver invoices in various different data formats, which was especially unappealing for SMEs with a low number of transactions (Self 2008). Therefore, the adoption rate remained at low levels and it was very difficult to convince smaller suppliers to participate in the solution. In the 2000s, B2B networks for e-invoicing began to simplify cooperation. Service providers established new ways of connecting suppliers and buyers by managing the conversion between differing standards of data exchange and other related services for either end of the process. However, while solutions today are available at nearly every level of sophistication, covering a variety of needs, the adoption rate is low and only increases slowly (Rombel 2007; Hernández-Ortega 2012; Koch 2017b).

From a legal perspective, early adopters were not able to replace the established paperbound process by utilizing EDI, since there was no legal framework allowing electronic documents to be deemed relevant for tax and reporting purposes. Paper held special significance as it represented material evidence (Tanner and Wölfle 2011). Starting around 2002, legislation in the EU and Switzerland began to remove the hurdles for electronic invoicing and allowed invoicing in a purely digital format for governmental tax audits. Thus, the legal basis was created to effectively handle electronic invoices and theoretically stop using paper. While this bold move should have made e-invoicing more attractive, especially to larger companies receiving thousands of invoices every month, e-invoicing was still not broadly adopted. This was due to differing standards among countries, complicating the process for companies engaged in cross-border business (Lejeune et al. 2003; Foryszewski 2006). More recently, governments reacted to the market need for standardization and less rigid legal frameworks (Koch 2017b). For example, in March 2016, the European Committee for Standardization (CEN) approved the first steps towards a unified European e-invoicing standard (de Jong 2017). In Switzerland, in September 2016 the obligation to sign electronic invoices digitally in order to prove authenticity and integrity during tax audits was eliminated (Eidgenössische Steuerverwaltung ESTV 2016).

Summing up, establishing e-invoicing makes good business sense for companies and governments. Combined political efforts translated into national law emphasized the interest of governments to promote e-invoicing. In the near future, all public administrative bodies in the EU member states have to accept electronic invoices in at least two standards, namely UBL and UN/CEFACT XML (Koch 2017b). Furthermore, electronic invoicing is requested from suppliers under certain circumstances, which will put pressure on organizations to implement e-invoicing.

3 Today's Obstacles

Since the driving forces behind the development in B2B relationships are mainly the companies receiving invoices from a multitude of different suppliers, the perspective of the invoice-receiving parties seeking to onboard their business partners is focused on hereinafter.

In the previous sections, we saw that many of the obstacles hindering the adoption of e-invoicing in B2B settings were already addressed and removed, especially from legal and technological perspectives. Nevertheless, why is it that e-invoicing is still not rising dramatically, although it would make good business sense to adopt it? Following the activities represented in boxes one to four in Fig. 1, we will look at the obstacles impeding the adoption of e-invoicing in organizations today.

Concerning activity one, *learn from best practices and seek advice*, rather than following a trend, this activity is about setting the stage for making an informed decision. Too often, decision makers do not have relevant and reliable information for their specific industry, do not interpret information correctly, or rely on outdated sources or even mere prejudices. Moreover, due to the fragmented market, various solutions are on offer that are not transparent regarding costs and benefits. In addition, not knowing about the needs and digital maturity of business partners will lead to poor decisions. Ongoing efforts to standardize and progress technologically could also impede decision-making, as it never seems to be the right moment to start implementation.

Concerning activity two, *assess skills and readiness and manage change*, besides other change initiatives that keep organizations occupied nowadays, restructuring the invoicing process is not likely to be given priority. Since the average paper-based invoice handling in organization typically shares an abundance of interfaces with related upstream and downstream business processes, automation is not established in passing. One of the biggest challenges in the internal sphere is to assess the organizations' culture concerning change and innovation. Furthermore, it would need to be decided what activities the organization can and wants to perform internally and what should be outsourced. Therefore, a critical assessment of the technological maturity of the organization is important, as well as connecting to relevant internal stakeholders such as IT, finance, accounting, procurement and other departments.

Concerning activity three, *involve business partners*, the company would need to work with external parties such as their business partners and service providers. However, companies (or more concretely: people working in companies) are reluctant to engage in a change project that involves the need to convince their business partners to change. Findings indicate that it is the human factor that poses the biggest challenge (Rombel 2007; Tanner and Wölfle 2011; Billing 2012; Hernández-Ortega and Jimenez-Martinez 2013; Hornburg 2017). In general, larger companies typically exhibit a high level of complexity due to organizational inertia, fragmented non-transparent processes and unclear accountability. Another challenge is that the involved parties might have differing understandings of the process, objectives and priorities of the project, as well as not being familiar with the technological aspects due to limited experience. Often, large corporates receive invoices from a variety of SME suppliers. However, when invoice-receiving companies implement e-invoicing, research shows that the rate of connected business partners is at not more than 30% after several years (Koch 2017a) because SMEs are not interested in implementing sophisticated e-invoicing infrastructure only for one client. This is even more the case as invoicing typically represents a process that either side has already organized in a satisfactory way. Nevertheless, the chances are that some suppliers are not yet on board simply because they were never approached by their customers (Hornburg 2017).

Concerning activity four, *design and assess solutions*, even if an e-invoicing process is already in place and standards are available, they are often not efficiently used to their full potential. This is due to the adoption of standards to "house rules" while at the same time proper documentation is lacking. This often happens when companies try to tweak an e-invoicing process to reflect company-specific needs (Tanner and Wölfle 2011). In addition, solutions might have been implemented as a quick fix, but are not accepted by the users and do not blend into the surrounding IT landscape. Furthermore, insufficient assessment of the stakeholders' needs can hinder business partners in joining the e-invoicing journey as they feel that either the solution is too sophisticated, complex and expensive for them to implement or too simple, thus not offering a compelling opportunity to automate.

The aforementioned points are reflected in data collected by swissDIGIN in 2016, asking a non-representative sample (n = 196) of the Swiss e-invoicing community about their primary concerns. The results show that across all company sizes, the top three issues reported are lack of expertise, high costs of implementation and the general inertia of the organization. Especially companies with more than 250 FTE (full-time equivalents) report lack of compatibility with existing systems as an additional obstacle. However, internal capacities and questions of IT security are less of a concern. Interestingly, it is mostly the invoice sending parties who report general inertia as a key issue, compared to the invoice receiving parties. This might be explained by the fact that it is typically the companies receiving invoices from an abundance of different suppliers that benefit the most from the implementation of e-invoicing solutions. This being said, the strong business case might offset inertia or set a stronger incentive to make an effort to overcome inertia.

4   Key Success Factors

After having identified the obstacles for e-invoicing today, the question remains of what can be done to drive implementation and reap the expected benefits. Again, the success factors are clustered following the activities represented in boxes one to four in Fig. 1.

Concerning activity one, *learn from best practices and seek advice*, it is crucial to seek information from trusted, reliable and non-opinionated sources. Considering today's availability of solutions for every level of sophistication, from e-mailing PDF documents to fully-fledged EDI-solutions, the definition of the specific objectives will determine what is and is not a good choice. Ideally, there is an opportunity to obtain recommendations and best practices from business partners in a comparable industry and combine them with on-site visits. Events such as expert conferences are also a good source of information and networking contacts. Collaborating with a service provider who has expert knowledge in the given industry can help to navigate the possibilities.

Concerning activity two, *assess skills and readiness and manage change*, experience from the swissDIGIN community shows that project management remains an important skill for e-invoicing. Practitioners report that e-invoicing projects too often lack proper project plans, resources and the definition of SMART (specific, measurable, achievable, realistic and timely) objectives being well matched with the other parties involved. From an organizational perspective, this also includes analyzing the current invoicing process critically. Too often, projects start with a picture of the current process as it should be rather than as it is. From the day the process was set up and documented, workarounds and tweaks might have emerged that never made it into the records. In the same vein, e-invoicing has to be approached as a holistic procure-to-pay exercise, involving representatives from the procurement, finance, logistics, tax and accounts-payable teams as well as the IT experts, in order to create a seamless workflow and obtain buy-in for the initiative (Taylor 2011; Billing 2012). Developing solutions together with those who work with the process every day will provide valuable insights and an opportunity to develop project- and change-management skills internally. Furthermore, accountability for the rollout and process ownership should be defined beyond the project end. Furthermore, it is key to gain commitment for the new solution, from top management to employees and all involved stakeholders. Therefore, classic change management guidelines should not be neglected, as change is likely to be met with resistance. The success of the project should not be jeopardized by not taking the human factor of change initiatives into account.

Concerning activity three, *involve business partners,* it is vital to evaluate which solutions would not only suit the needs of one's own organization, but also those

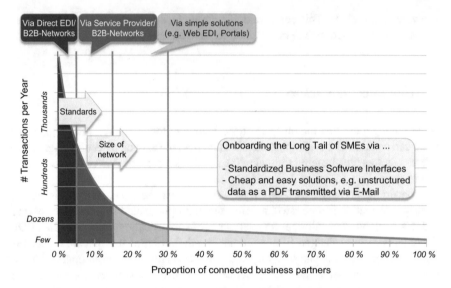

Fig. 2 Onboarding the long tail of SMEs (based on Tanner and Wölfle 2011)

of the business partners. Figure 2 shows the proportion of the connected business partners together with the number of transactions per year.

While those partners with high numbers of transactions are relatively easy to onboard onto e-invoicing through direct EDI or through EDI by using B2B networks run by service providers, the remaining long tail of SMEs has yet to be convinced (Tanner and Wölfle 2011). Political efforts therefore focus on the onboarding of the long tail of SMEs (Tanner and Wölfle 2011; Taylor 2011). For a company, it is crucial to proactively involve business partners to exploit the full potential of e-invoicing. That having been said, it is worthwhile to conduct a thorough analysis of the willingness and technological maturity of the company's business partners in order to clarify what range of solution options a company should offer. For the rollout, it is advisable to prioritize business partners with high numbers of transactions and those who are experienced in e-invoicing. A mutually shared understanding across stakeholders and compelling benefits for the suppliers are additional key success factors for this activity.

Concerning activity four, *design and assess solutions*, it is important to ensure that there is a unified internal process in place, supplied with high-quality core data and working with a broadly accepted standard. Further key elements for the successful rollout are collaboration with a service provider and being connected to a relevant B2B network that offers attractive pricing models for the suppliers and facilitates an easy onboarding process. Moreover, it is worth considering the different options available between paper-based invoicing and a fully-fledged EDI solution to obtain the long tail shown in Fig. 2 on board and find the appropriate range of solutions for the specific company. Figure 3 shows different levels of sophistication in e-invoicing

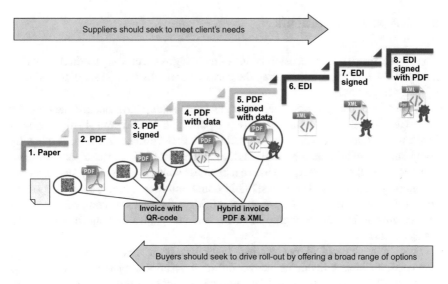

Fig. 3 Levels of sophistication in e-invoicing solutions (based on Tanner 2016)

solutions. Organizations can decide to approach the implementation of e-invoicing from either side of the model.

Starting with a paper-based process, one could decide to stop using paper through scanning, turning paper documents into PDFs. Since this is rather close to the original process and processing PDFs via e-mail is an established procedure in business settings, the effort needed to convince people to participate and further change management is small. However, the benefit is limited, as this approach does not allow for the automated processing of data. This is the basic approach mostly taken by suppliers needing to meet their client's requirements. Based on the client's needs, they will likely have to move up to a more sophisticated solution such as sending a digitally signed PDF invoice or structured data (EDI) at a certain point.

On the other hand, a company might decide to make a bold move and go for the fully-fledged EDI solution. This approach is more complex to implement and calls for proper project- and change management since it implies major changes compared to the original process. Furthermore, this approach requires trust in technology and automated controls. However, it also allows for fully automated processing of invoices while being highly scalable. This is the approach mostly taken by organizations receiving thousands of invoices from their suppliers. However, as they aim to onboard a maximum number of business partners, they tend to open up to simpler solutions eventually.

Either way, organizations pursuing e-invoicing will need to evaluate the individual requirements of their business partners in order to determine what range of solutions needs to be offered, depending on transaction volumes and technological maturity. They will eventually see the need to adapt to simpler or more sophisticated solutions in order to maximize the number of connected business partners.

5   Recent Developments

Concerning the EU's goal of establishing e-invoicing as the primary method of invoicing by 2020, it is worth considering the current developments and state of readiness of involved parties.

From a legal perspective, recent changes in legislation have opened the door for less sophisticated ways of e-invoicing (e.g. by omitting the obligation to digitally sign invoices). For instance, sending documents in a PDF format via e-mail is a legally compliant way of e-invoicing, as long as the authenticity and integrity of the document can be proven through adequate internal business controls and is duly supported by accounting documents. The swissDIGIN community regards this as an important step in driving e-invoicing in general, as it bridges the needs of companies seeking to maximize efficiency through e-invoicing and companies looking for simple and cheap solutions.

From a technological perspective, there has been a recent convergence of standards, making e-invoicing across different solutions, systems and standards easier. Furthermore, there is a trend towards business software solutions implementing e-invoicing as a basic functionality, following established standards. The incorporation into existing invoicing processes can be achieved with little effort, especially with options such as e-invoicing through PDF/A-3 documents enriched with attached structured data in UN/CEFACT XML format.

Opinions about PDF invoices were collected by swissDIGIN in 2016, surveying a non-representative sample of companies in the Swiss e-invoicing community mainly receiving invoices (n = 71) and mainly issuing invoices (n = 58). Results indicate that the majority of companies in either role support PDF invoices in general as contributing to the overall goal of widely adopting e-invoicing. However, 18% of the companies that mainly receive invoices argued that the acceptance of PDF invoices might have an adverse effect, because it hinders the adoption of fully automated EDI solutions.

Furthermore, current developments in Switzerland indicate that a new way of invoicing might soon enable even simpler solutions that would facilitate the onboarding of the long tail of suppliers. This new option is based on the standards of the Single Euro Payments Area (SEPA), which harmonizes bank transfers across 34 states in Europe, and was established using a unified data format following the ISO 20022 standard (The National Adherence Support Organisation (NASO) of Switzerland 2017; ISO-20022.CH). In 2017, the Swiss financial center presented its new QR-bill stating that "the new, future-oriented solution enables the different interest groups to meet the challenges of digitalization and regulation in an efficient way" (SIX Interbank Clearing Ltd 2017). The QR-bill will replace traditional payment slips from January 2019.

Since the QR-code will contain all necessary payment information, it can be used physically on paper or digitally, e.g. in a PDF and can be read automatically with a QR-code reading software. As shown in Fig. 4, the new QR-bill not only includes the

Fig. 4 New swiss payment slip with QR-code (PaymentStandards.CH 2017)

QR-code with a Swiss cross on the left, but also displays the payment information in text format on the right, making it readable for non-automated processing as well.

With the mandatory use of the new payment slip, invoice-processing software will inevitably have to incorporate relevant functionalities for generating invoices and reading the QR-bill. Furthermore, it will be possible to add reference information for invoice verification purposes. Therefore, Fig. 3 shows a QR-code symbol in the first three steps of the model. This is to symbolize that the QR-bill can serve as a simple measure transferring paper-based information into a machine-readable format allowing for a first level of automation. From the QR-bill, the next step of sophistication is the hybrid format combining a PDF/A document with structured data in UN/CEFACT XML format. This accommodates mainly small suppliers seeking to comply in a simple way with their business customer's requirements.

From an economic perspective, the size of B2B networks becomes the crucial factor determining the value and competitiveness for service providers. Thus, the market landscape of service providers started to consolidate as they sought to leverage synergies and market coverage to broaden their networks (Mayer 2011; Taylor 2011). Unsurprisingly, e-invoicing service providers running B2B networks supporting EDI also are not positive about the trend towards simple solutions such as PDF invoices, given that their business model today benefits from complexity, non-unified standards and a demand for digital signatures. Hence, in the swissDIGIN survey of 2016 service providers (n = 24) express little support for statements such as "PDF invoices without electronic signature should be accepted as evidence by the receiving party". Only 29% of the service providers state their full agreement with this statement, whereas 67% of the invoice sending companies and 46% of the companies mainly receiving invoices do.

6 Conclusion

After having looked at electronic invoicing as an example of the digitalization of a business process, what are the generalizable learnings and conclusions we can draw?

Concerning the PESTEL factors, it becomes obvious that the political, technological and legal infrastructure is ready to support digitalization and much effort has been put into facilitating the digital change. Furthermore, from an economic perspective, the business case for the digitalization of business processes is strong. This does not mean that all determining factors are indeed optimal, yet small efforts can already go a long way.

Referring to the activities of the company-internal change shown in Fig. 1, concerning activity one, *learning from best practices and seeking advice*, the mere news of good general conditions for digitalized processes will not make companies change. Therefore, the digitalization of business processes should be understood as a core task of today's division managers who need to stay abreast of the latest developments in their industry, as they are the designers of processes and systems. Since there will never be the ideal moment where there is absolute security regarding legal and technological circumstances, company culture should encourage the taking of a controlled risk and the building of adequate failure tolerance.

This is especially true when it comes to the second activity, *assess skills and readiness and manage change*. Managing change internally is known to be challenging and has been examined in countless studies and models by academics and practitioners. From the example of e-invoicing it can be derived that a culture of change and innovation has to be developed for both top-down and bottom-up initiatives. Employees should not only be involved as the experts for their processes, but they also should be further developed through designing, piloting and documenting the project. Thus, knowledge is built and retained within the company.

For the third activity, *involving business partners*, collaboration in and with B2B networks is the crucial success factor concerning standardization and meeting legal requirements. Any digitalization effort is likely to fail or never reach its projected potential if key trade partners are not on board. Companies need to collaborate with their business partners and design solutions reflecting the willingness to digitalize and the digital maturity of both parties in a joint effort. Of course, cost-benefit considerations have to be acceptable for both parties.

The fourth activity, *design and assess solutions*, takes the point made for activity three even further. Starting with a viable solution for one's own situation is good. However, this might not satisfy the needs of business partners. Therefore, companies aiming to digitalize their B2B processes should seek to adapt to their partner's needs by providing a range of solutions that suits both larger corporates and small businesses, depending on whom they do business with. Ideally, solutions would have a modular design that is scalable, allowing for extensions concerning the processes, geographical area, functions and organizational units. This would allow solutions to grow alongside the needs of the organization and its business partners. Standards should be followed and rigorously adhered to, as this will contribute to a smooth rollout. Digital B2B networks are by far the predominant solution scenario when it comes to integrating the processes and systems with a wide range of business partners.

Summing up, it can be concluded from the example that, in order to digitalize processes successfully, organizations need to broaden their perspective. Instead of initiating an effort focusing on how to turn the exact current process into a digital version, businesses should assess and understand the PESTEL factors at play as well as the four activities from learning from best practices to designing and assessing appropriate solutions. However, we recognize that there is a range of individual cases and the results of our research and experience incorporated in this article are only generalizable to a limited extent. Furthermore, as much as there is no "one size fits all"-solution, there is also none that will last forever. Rather, as the business environment changes and technology matures, there will be a need to re-assess processes and solutions from time to time. Most importantly, the human factor of change cannot be underestimated. Besides standard change management practices, companies should seize the opportunity to further develop the people in the organization through the digitalization effort. Acquiring project management skills, expert knowledge and relevant experience in innovating business processes will serve as an invaluable asset in the long term.

References

Airey K (2014) Basware Predicts 2014 will be the "Year of E-Engagement" for B2B commerce, Driven by Explosive Growth of Social, Mobile and Cloud. Investment Weekly News, Atlanta

Billing N (2012) E-Invoicing: overcoming the challenges of implementation. Supply Chain Europe; Leatherhead 21, pp 10–11

de Jong F (2017) EU E-invoicing standard EN 16931 unanimously approved. http://eeiplatform. com/18364/european-e-invoicing-standard-unanomous-approval/. Accessed 4 Apr 2017

Eidgenössische Steuerverwaltung ESTV (2016) Der elektronische Geschäftsverkehr EGV. https://www.estv.admin.ch/estv/de/home/mehrwertsteuer/fachinformationen/elektronischer-geschaeftsverkehr.html. Accessed 28 Jun 2017

European Commission (2013) E-invoicing in public procurement: another step towards end-to-end e-procurement and e-government in Europe. http://europa.eu/rapid/press-release_IP-13-608_en. htm. Accessed 3 Apr 2017

Foryszewski S (2006) The evolution of B2B electronic invoicing. Credit Control, Hutton, vol 27, pp 35–38

Hawser A (2009) Paperless payments take off. Global Finance, New York, vol 23, pp S8–S9

Hayward A (2013) E-invoicing in the construction industry: e-Invoicing holds the key to improving accounts payable efficiency and maximizing supplier relations. Credit Control, Hutton, vol 34, pp 27–30

Hernández-Ortega B (2012) Key factors for the adoption and subsequent use of E-invoicing. Academia, Bogotá, vol 15

Hernández-Ortega B, Jimenez-Martinez J (2013) Performance of e-invoicing in Spanish firms. In: Information Systems and eBusiness Management. Heidelberg, vol 11, pp 457–480. http://dx.doi. org/10.1007/s10257-012-0203-y

Hornburg M (2017) E-invoicing: why aren't we there yet? http://www.coupa.com/blog/e-invoicing-why-aren-t-we-there-yet? Accessed 4 Apr 2017

ISO-20022.CH ISO 20022 Schweiz. https://www.iso-20022.ch/lexikon/iso-20022/. Accessed 3 May 2017

Kioses E, Pramatari K, Doukidis G, Bardaki C (2007) Measuring the business value of electronic supply chain collaboration: the case of electronic invoicing. In: BLED 2007 Proceedings, vol 53

Koch B (2017a) Business Case E-Invoicing/E-Billing. http://www.billentis.com/e-invoicing-businesscase.pdf. Accessed 17 May 2017

Koch B (2017b) E-Invoicing/E-Billing—significant market transition lies ahead. Billentis, Wil, Switzerland

Lejeune I, Korf R, Grunauer A (2003) New EU e-commerce directives: a move towards e-Europe? J Int Taxat Boston 14:12–23

Mayer S (2011) E-Invoicing/E-Billing—from evolution to revolution. PR Newswire Europe Including UK Disclose, New York

Pan-European Public Procurement OnLine (2017) What is PEPPOL. http://peppol.eu/what-is-peppol/. Accessed 3 May 2017

PaymentStandards.CH (2017) The financial center presents the QR-bill. https://www.paymentstandards.ch/en/shared/news/2017/qr-bill.html. Accessed 3 May 2017

Rombel A (2007) Paving the way for E-invoices. Global Finance, New York, vol 21, pp 21–22

Self E (2008) Understanding the trend toward E-invoicing networks. Business Credit, New York, vol 110, pp 36–37

SIX Interbank Clearing Ltd (2017) Payment system harmonization—from January 2019 the current Swiss payment slips are being replaced by the QR-bill. https://www.paymentstandards.ch/dam/downloads/media-release-qr-bill-en.pdf. Accessed 3 May 2017

Tanner C (2016) PDF-Rechnungen, aber richtig! In welchen Szenarien die PDF-Rechnung passt. http://www.swissdigin.ch/ver_161123. Accessed 19 Apr 2017

Tanner C, Wölfle R (2011) E-Invoicing—Elektronischer Rechnungsaustausch, 1st edn. edition gesowip, Basel, Switzerland

Taylor N (2011) All your eggs in one basket? Supply chain Europe. Leatherhead 20(34):36

Tenhunen M, Penttinen E (2010) Assessing the carbon footprint of paper vs. electronic invoicing. In: ACIS 2010 proceedings

The National Adherence Support Organisation (NASO) of Switzerland (2017) Single Euro Payments Area (SEPA). https://www.sepa.ch/en/home/sepa.html. Accessed 3 May 2017

Ulrich J (2002) Payables automation and electronic settlement: the time is now. AFP Exchange; Bethesda, vol 22, pp 44–46

Wölfle R (2000) E-Business erfolgreich planen und realisieren. In: Schubert P, Wölfle R (eds) E-Business erfolgreich planen und realisieren: Case Studies von zukunftsorientierten Unternehmen. Carl Hanser Verlag, München, Wien, pp 13–26

Marketing Automation

A Project Framework in Support of Digital Transformation

Martina Dalla Vecchia and Marc K. Peter

Abstract While the benefits of marketing technologies in organizations were already being discussed in the 1960s, the contemporary approach to Marketing Automation has only been in existence for a few years. This chapter addresses the digital opportunities in marketing as an important component of an organization's digital transformation initiative. Driven by various business challenges centered around the organization's customers and market channels, the goal of this literature review is to provide an overview of the potential applications of Marketing Automation and the growing market of available software solutions. A particular focus is set on a pragmatic best practice framework for Marketing Automation that includes all major components, ranging from strategic considerations, customer journey analysis to use cases along with data management, content marketing and channel management. While there are some simple implementation opportunities, a best practice framework will support an organization to achieve its ambitious goals. However, the opportunities for Marketing Automation are still evolving as new tools and channels come on to the market.

Keywords Digital marketing · Marketing automation · Digital transformation
Marketing communications · CRM · Content marketing · Digital leadership
in marketing

M. D. Vecchia (✉)
Institute for Information Systems, University of Applied Sciences and Arts Northwestern
Switzerland, Peter Merian-Strasse 86, 4002 Basel, Switzerland
e-mail: martina.dallavecchia@fhnw.ch

M. K. Peter
Institute for Competitiveness and Communication, University of Applied Sciences and Arts
Northwestern Switzerland, Riggenbachstrasse 16, 4600 Olten, Switzerland
e-mail: marc.peter@fhnw.ch

1 Introduction

1.1 Background and Definition

Digital transformation is driven by new means of technology and communication, which influence society, work and all areas of life. In turn, transformational initiatives lead to new products, services and business models (BMWi 2015; SECO 2017). It is regarded as a prerequisite for organizations to have access to transparent and pragmatic frameworks in order to successfully drive and support a digital transformation initiative (Schwaferts 2016).

In the fields of marketing, sales and customer service, a Customer Relationship Management (CRM) solution is the highest prioritized project with regard to digital transformation, and e-mail marketing is defined as the most important tool to attract new customers (Peter 2017). Both form part of Marketing Automation, marketing's answer to the digital transformation. The basic idea is to design repetitive processes in such a way that they run automatically but are tailored to each individual customer. In this sense, Marketing Automation is a form of mass customization for branding, communication, social media and sales (Schoepf 2017).

The concept of Marketing Automation evolved in the sixties with the rise of technology. It was stated that automation "…can help in the gathering and analysis of facts" (Head 1960, p. 35) and therefore enables automated audits and forecasts, and facilitates transactions. The focus on data-based decision-making in marketing is a core component of today's Marketing Automation, while the second aspect, to support platforms and communication channels, has not yet been discussed. In 1962, Christian (p. 81) described how in external communication, the computer would make its presence felt by means of "…analyzing the relationships between advertising and sales dollars". However, despite all of the advantages of the existing automation solutions, there is still a need for good planning which takes into account the customers' needs and is carried out by well-trained specialists (Goeldner 1962).

In the nineties and at the beginning of the new millennium, the focus has shifted from data-based decision-making to IT (information technology) supported approaches, such as paperless marketing and sales departments and enterprise marketing automation (Flanagan 1995; Doyle and Georghiou 2001). This enhanced view of Marketing Automation included workflow support, behavioural analysis, decision support, content and channel management. Here, clear business goals, the right skills, IT and process management were already regarded as success factors for a "balanced transformation" (Doyle and Georghiou 2001, pp. 177/178).

In order to automate processes in marketing, three important components are required (Schoepf 2017):

1. Detailed knowledge about the customer (the customer journey);
2. Data/information (CRM, purchasing history, user behavior, etc.); and
3. Marketing Automation software.

Therefore, it is not surprising that Marketing Automation originated in the field of e-mail marketing. The marketing teams know the journey of their customers: They know when to send an e-mail to initiate a purchase or when to send a follow-up email as part of the customer service. In addition, they have a customer database that is, in the ideal scenario, connected to the CRM system or already fully integrated. In the most straightforward scenario, one could start a simple process automation with a standard e-mail solution. Years ago, newsletter software providers recognized this market potential and trimmed their solutions to automation. Today, almost all newsletter programs offer the possibility of automating simple processes. Other providers in the digital marketing environment followed (SiriusDecisions 2017).

1.2 Business Challenges

While Marketing Automation provides multiple advantages combining many marketing disciplines (e.g. data-driven decision-making, content and platform management, and customer communication) in an automated, process-driven framework (Doyle and Georghiou 2001; Head 1960; Järvinen and Taiminen 2016; Schoepf 2017; Todor 2016), organizations that want to implement a Marketing Automation project are also faced with multiple business challenges.

Digital customer behavior has changed drastically over the last years: Mobile Internet usage prevails in almost all customer segments and the use of social media, as well as the intensive use of search engines, has massively changed consumer behavior and brand perception (Schoepf 2017).

According to the literature, the four major challenges for businesses with regard to Marketing Automation in the wider sense are:

1. Customer journey: Consumers expect organizations to provide a professional customer experience that is suited to their individual situation and needs, i.e. a personalized customer journey (Foroudi et al. 2016; Joshi 2014);
2. Personalization: Businesses should recognize the consumer's drivers and needs in order to enhance services, channels and communication platforms with personalized content, which is aligned with the above customer journey (Luke 2013; Riecken 2000);
3. Multi-channel management: Businesses need to align their channels, provide multiple channels based on the needs of the consumers and align their value proposition and brand message while synchronously controlling all channels (Batra and Keller 2016; Chen et al. 2014); and
4. Data management: Current data for customer interactions and the legally compliant handling of customer data is a fundamental question for businesses. In addition, it is necessary to establish a clear governance for data management and the resulting business actions (Levy 2015; Moorthy et al. 2015).

Therefore, every business should know its customers' journeys based on the most important target groups (personas), provide a holistic view of all channels and have them aligned to the personas while maintaining an adequate level of personalization, and ensure that data management functions and processes are in place.

1.3   Research Goal and Methodology

Given the various business challenges with regard to the adoption of marketing technologies (focus Marketing Automation), the goal of this book chapter is to provide an overview of the application scenarios for Marketing Automation (Sect. 2) and the current software solutions available to organizations (Sect. 3).

Most importantly, in the context of digital transformation and the need for transparent and pragmatic frameworks, a best practice project framework for Marketing Automation will provide an overview of the key components of the Marketing Automation. With such a framework, organizations will be able to identify important work streams or sub-projects, build a roadmap and prepare for successful project implementation.

The choice in research methodology depends on various factors. While first references to Marketing Automation (or more precise, marketing technology) could be found in the literature of the 1960s, it is only in the past ten years that the discipline has been known to a wider group of researchers. Therefore, the field is relatively young and the amount of scientific literature is manageable. Since qualitative research methodologies are gaining in prominence in management research (Singh 2015), a literature review was selected as the most appropriate methodology for a review and synthesis of existing knowledge.

The review was conducted following steps recommended by Levy and Ellis (2006): Firstly, following the search strategy and keyword definition, the literature was gathered and screened (via literature databases and electronic resources); secondly, the literature was processed and classified; and finally, a report was produced. As part of the second step, Levy and Ellis (2006, p. 182) suggest that literature needs to be understood, analyzed, synthesized and evaluated. The authors applied this process and ran a second iteration to ensure that the relevant knowledge was gathered.

The following three sections will provide an overview of business cases, the available software solutions and the best practice project framework.

2   Business Cases for Marketing Automation

In general, Marketing Automation can be used for three key business cases:

1. Internal communication
2. External corporate communications
3. Marketing and sales communications.

In internal communication (Kang and Sung 2017; Mac and Shirley 2015), automation ensures effective communication between the organization, or rather the management, and the employees. Especially with regard to the many organizational transformation and change programs of today, automation is increasingly being used for project support. The solutions that are present in the market offer newsletter functions, notifications and surveys with many options for usage analysis.

In the case of external corporate communication (Cornelissen et al. 2001; Ki and Ye 2017), automation supports the systematic planning, implementation and control of all digital communication channels in order to provide the various stakeholder groups with information. Especially with the growth of social media, automation can create a great deal of added value. Standard functions for external communication are the identification of relevant contacts (e.g. journalists, bloggers), the organization of media lists, the creation of news pages or the administration of media centers, the coordination and analysis of e-mail messages and invitations, the search for relevant content and reports, and the publication of communications on social media.

Finally, in the case of marketing and sales communication (Balmer and Yen 2017; Batra and Keller 2016; Luxton et al. 2015), automation is used for many tasks and thus provides controlled customer management and support. The wide-ranging fields of application cover classical marketing communication, customer service, enable direct sales ("real-time buying") or the acquisition and retention of potential new customers. In addition, target group identification, campaign testing, mobile marketing and the coordination of campaigns on social media channels and via e-mail are supported.

While Marketing Automation will be able to support all three business cases, not all features of the available software solutions will be meaningful or required.

3 Marketing Automation Software Solutions

Over the past years, many providers of Marketing Automation solutions have emerged. Therefore, the software evaluation is very complex. Depending on the business case (see Sect. 2) and specialization of the software solutions provider, there is most likely more than one provider, who will fulfill the organization's requirements. According to Schoepf (2017) and SiriusDecisions (2017), today most automation solutions offer:

- Form and landing page designer
- E-mail designer
- List management for customers and interested parties
- Web analytics/reporting for e-mail and landing page performance
- Support .

Advanced features include:

- Workflow designer
- A/B testing
- Social media integration
- E-commerce support (e.g. contact shopping basket dropouts)
- CRM interface
- Text messages (SMS)
- Interface to affiliate platforms
- Outsourced call center
- Helpdesk functions.

The Marketing Automation market ranges from open source to full service providers. The 2017 market overview of SiriusDecisions (2017) analyses the leading software solutions for SMB (Small and Medium sized Businesses) and Enterprise clients against nine use cases with relevant criteria. Sitecore, an Australian provider, has mainly large enterprise customers, while the other solution providers, such as IBM, Microsoft Dynamics Marketing, Oracle Eloqua etc., have a mix of enterprise and SMB clients.

The online platform G2 Crowd (2017) provides a free, interactive guide based on user ratings for Marketing Automation solutions. In terms of user satisfaction and market presence (i.e. market share), HubSpot®, Marketo® and Pardot® are the leading solutions. These providers are positioned in the SMB market and subscription prices start at around US\$200/month for the basic version. They offer a comprehensive solution package with an attractive price-performance ratio.

For sophisticated, consulting-led engagements with the larger solution providers such as Adobe®, Oracle® and IBM®, projects are more complex and implementation time frames are generally longer in order to fulfill the client's requirements.

Regardless of project and/or client size, a basic set of components must be considered in order to successfully implement a Marketing Automation solution. These components will be presented in Sect. 4 as part of the Marketing Automation project framework.

4 Marketing Automation Project Framework

As discussed in Sect. 1, Marketing Automation provides many advantages but businesses are faced with challenges such as customer expectations, multi-channel alignment, personalization requirements and data management. Due to this complexity, their importance for the organization and high investment levels, projects should also be well planned and follow a best-practice approach.

Based on the literature review, the major components of Marketing Automation projects have been identified and summarized in ten framework components:

1. Strategy and project goals
2. Customer activity analysis/customer journey
3. Use cases and customer scenarios
4. Segmentation and data management
5. Predictive intelligence
6. Content marketing
7. Channel management
8. Analysis and control
9. Technical infrastructure and processes
10. Project management

The following sections will provide an explanation of the various components.

4.1 Strategy and Project Goals

The definition of the strategy and project goals at the beginning of the project is a key step in order to formulate the desired qualitative as well as quantitative results. Research shows that, due to the initial definition of the strategy and goals, managers of software projects will provide the appropriate resources accordingly (Abdel-Hamid et al. 1999).

The following elements need to be defined and articulated around project strategy and goals:

- Overview of the challenges to be solved by Marketing Automation as well as possible strategic approaches
- Concrete objectives regarding new marketing skills and financial success metrics, e.g.:

  - Acquisition of new customers
  - Customer loyalty and customer-centric goals based on the customer lifecycle and journey
  - Customer recovery
  - Better coordination between sales and marketing
  - Customer service

- Project framework components such as priorities, available and/or required resources, deadlines and budgets.

4.2 Customer Activity Analysis/Customer Journey

In the second step, a marketing analysis is carried out which identifies strategic customer information and requirements in order to define possible use cases (Bucklin and Sismeiro 2009; Leeflang et al. 2014). The entire automation approach is based on a customer and data-driven strategy to ensure a positive customer experience across the customer journey (Borges Tiago and Verissimo 2014; Leeflang et al. 2014). Here, the following activities are carried out:

- Analysis and definition of the critical customer experience components, i.e. the customer experience;
- Identification of sources and methods to gain customer feedback;
- Analysis of customer feedback, customer opinions and definition of typical problem areas or optimization potential; and
- Holistic analysis (actual state) and redefinition (target state) of the customer activity along the customer journey, business value chain and all channels.

Personas are often used in this context (Dion and Arnould 2016; Pruitt and Adlin 2010): Marketing personas are fictitious individuals who reflect a specific target group with their individual needs regarding the customer experience. A persona includes typical features such as the person's name and job title, a photo, demographic characteristics, problems and needs, and a typical daily routine.

4.3   Use Cases and Customer Scenarios

Based on the findings of the customer activity analysis, existing data and the defined strategy, the most important use cases or customer scenarios that are to be automated are defined (Li et al. 2011; Longo and Vilain 2015). It is worthwhile to prioritize and reduce complexity to ensure a high level of implementation success. With experience gained from initial testing, simple use cases can be further enriched and optimized.
Examples of use cases are:

- Based on certain variables, a potential new customer is assigned to a specific field employee;
- After 30 days of inactivity, an e-mail will be sent with product tips; after 50 days of inactivity, a call from a customer service agent will be actioned;
- Due to the first website visit and the first specific purchase, a customer is assigned to a certain campaign; and
- Due to his or her online behaviour and persona, a new customer receives pre-defined e-mails.

4.4   Segmentation and Data Management

The various segments or target groups are described in more detail. The central data criteria, which then control the automation, are defined for each segment and/or persona.

In Marketing Automation, analytical marketing, together with segmentation, is employed to control communication (Bucklin and Sismeiro 2009; Li et al. 2011; Schäfer and Kummer 2013), i.e. the generic communication objectives of marketing recede into the background. In many cases, this will reduce the number of receivers but will increase the quality of the communication or the response rate for campaigns, especially for lead generation and nurturing. This is, among other things, also influenced by the personalization of the content (Li et al. 2011).

The requirements for data management are also defined, such as the data quality (and the subsequent work to improve the data) as well as the requirements for the customer databases (CRM systems) and their analytical capabilities (Gerrikagoitia et al. 2015; Kwan et al. 2005).

4.5 Predictive Intelligence

For more complex automation projects or for organizations with layered multi-channel customer interactions, quantitative models are developed to calculate the strength of the customer relationship and their future potential (Borges Tiago and Verissimo 2014; Dueñas-Fernández and L'Huillier 2014; Martens et al. 2016). The probability of a customer activity (propensity score) and the intensity (engagement score) of the customer relationship/customer interaction are calculated:

- The "propensity score" is calculated on the basis of the information collected on demography and online behavior (potentially enriched with offline data); and
- The "engagement score" is calculated on the basis of the totality of all interactions along the customer process. This includes the use of the website, visits to business premises/retail outlets, brand engagement on social media, telephone calls, activities on social media, etc.

In both models, the individual customer data is compared to the total data of all customer movements/customer information, i.e. the system learns over time and thus automation is optimized.

4.6 Content Marketing

Content marketing aims to generate, manage and distribute target group relevant content, created according to journalistic criteria (such as texts, images, graphics, interactive, audio, videos, blogs) in a media-appropriate manner across all communication channels in order to address the internal and external target groups more successfully (An Kee and Yazdanifard 2015; Järvinen and Taiminen 2016).

As the content and therefore content marketing play a central role in Marketing Automation, the three sub-areas of (1) the content audit, (2) the content strategy and (3) content creation and distribution must be addressed in detail (Borges Tiago and Verissimo 2014; Schäfer and Kummer 2013):

- Content audit:

- Identification and description of the customers' need for information
- Analysis of the content, activities and channels used by the competition
- Identification, rating and prioritization of existing content and content gaps

- Content strategy:

- Strategy definition including core themes and media (e.g. text, video, audio, image)
- Adaptation and/or coordination with higher-level marketing objectives and the Marketing Automation strategy
- Definition of the content framework and requirements for the editorial team
- Definition of the first automation campaigns

- Content creation and distribution:

- Definition of the necessary infrastructure, including Content Management Systems (CMS), content lifecycle management, processes and roles
- Definition of the most important channels, including definition of channel rules (see Sect. 4.7)

4.7   Channel Management

The digital and physical channels of marketing, sales and customer service must be aligned to achieve both corporate objectives and a positive customer experience (Lichtenthal and Eliaz 2003; Schäfer and Kummer 2013). An important objective is the ability to communicate with customers in new forms through automation and to nurture them across the various channels (Leeflang et al. 2014).

The channels are defined and described with their relevant priorities, their frequency for customer interaction and their respective use cases based on the marketing strategy, the customer processes and the objectives of content marketing. The question that needs to be answered is how the current channels must be used and/or modified and which channels are missing and therefore need to be developed. In most cases, a matrix is used to coordinate segments or personas, channels and customer interfaces ("touch points"), use cases and campaigns to achieve the qualitative and quantitative marketing objectives of the automation.

4.8   Analysis and Control

Digital channels and interactions enable the Marketing Automation solution to constantly measure the efficiencies of marketing investments and campaigns. With the test results (for instance from A/B testing) and data collections based on the various customer interactions, a level of optimization is sought which will improve future marketing projects and campaigns (Leeflang et al. 2014; Li et al. 2011).

Accordingly, the automation project will define which systems are used for analysis and control and which quality processes are to be implemented. The focus is on quantitative methods and systems that measure the critical success factors before and after the campaigns. In real-time or following a completed campaign, findings will be implemented to optimize the campaign. This creates a living system that achieves enormous marketing efficiencies.

4.9 Technical Infrastructure and Processes

The framework components and requirements for the infrastructure as well as the processes are defined in detail (Borges Tiago and Verissimo 2014; Wang et al. 2013). In addition to the Marketing Automation software solution, this will include the channels (website, e-mail, telephone marketing, etc.), the database solutions (CRM and CMS) and ERP (Enterprise Resource Planning) system that covers product information, order fulfillment, accounting etc. required for operational purposes.

Many marketing, sales and customer service processes (which may also include company-wide processes) are affected by the automation, which drives process modifications. Specifically, the interfaces between marketing, sales and customer service will be modified (automated or optimized) which in turn, in most cases, also leads to structural changes in these departments.

4.10 Project Management

Lastly, the traditional tasks of project management are applied (McAlister 2006). These include:

- The resources and roles necessary for the introduction of Marketing Automation and ongoing operations;
- The identification of the new skills that are necessary and the planning of the training activities based on these needs; and
- A project schedule with milestones, a budget and the success metrics for project control following implementation.

Through good planning, oriented towards the business objectives and the market, i.e. the customer requirements, Marketing Automation can be used profitably for various business cases. It will strengthen the customer experience and at the same time, drive growth for the business.

5 Summary and Conclusion

Driven by the broader agenda of digital transformation, Marketing Automation combines a series of marketing technologies, including CRM and e-mail marketing, to design and implement automated processes based on data analytics that trigger use cases and orchestrate campaign plans, content assets and channel execution. The primary challenges for organizations in implementing such a solution (with many providers to choose from) are to define a customer journey that can be automated, yet personalized based on rich data and analytics, and to manage all channels simultaneously. Successful implementation that utilizes the Marketing Automation project framework will provide automation for internal communication, external corporate communications as well as marketing, sales and customer service communication.

As there is literature available that permits the definition of the best practice framework, there are a number of new tools and channels emerging that still need to be understood and their potential in a Marketing Automation software solution defined, including apps, chatbots and messaging services. As a rule, simple processes for new customer acquisition and customer activation can be implemented with basic automation software. The use of a Marketing Automation solution is associated with greater effort, as there are various processes to design and implement. In some cases, this leads to structural changes in the marketing organization. This requires a more detailed examination of the project scope, maintenance implications, and running costs compared to the expected yield. Only when the organization is ready for digital transformation in marketing, can the investment for the development of process optimization and a better customer experience be undertaken.

References

Abdel-Hamid T, Sengupta K, Swett C (1999) The impact of goals on software project management: an experimental investigation. MIS Q 23:531–555. https://doi.org/10.2307/249488

Balmer J, Yen D (2017) The internet of total corporate communications, quaternary corporate communications and the corporate marketing internet revolution. J Mark Manag 33:131–144. https://doi.org/10.1080/0267257x.2016.1255440

Batra R, Keller K (2016) Integrating marketing communications: new findings, new lessons, and new ideas. J Mark 80:122–145. https://doi.org/10.1509/jm.15.0419

Bucklin R, Sismeiro C (2009) Click here for internet insight: advances in clickstream data analysis in marketing. J Interact Mark 23:35–48. https://doi.org/10.1016/j.intmar.2008.10.004

Bundesministerium für Wirtschaft und Energie (2015) Industrie 4.0 und Digitale Wirtschaft - Impulse für Wachstum, Beschäftigung und Innovation. Bundesministerium für Wirtschaft und Energie, Berlin

Chen K, Kou G, Shang J (2014) An analytic decision making framework to evaluate multiple marketing channels. Ind Mark Manage 43:1420–1434. https://doi.org/10.1016/j.indmarman.2014.06.011

Christian R (1962) The computer and the marketing man. J Mark 26:79–82. https://doi.org/10.2307/1248310

Cornelissen J, Lock A, Gardner H (2001) The organisation of external communication disciplines: an integrative framework of dimensions and determinants. Int J Advert 20:67–88. https://doi.org/10.1080/02650487.2001.11104877

Dion D, Arnould E (2016) Persona-fied brands: managing branded persons through persona. J Mark Manag 32:121–148. https://doi.org/10.1080/0267257x.2015.1096818

Doyle S, Georghiou J (2001) Software review: a process change model to meet the Enterprise Marketing Automation (EMA) vision. J Database Mark Cust Strategy Manag 8:176–182. https://doi.org/10.1057/palgrave.jdm.3240033

Dueñas-Fernández R, Velásquez J, L'Huillier G (2014) Detecting trends on the web: a multidisciplinary approach. Inf Fusion 20:129–135. https://doi.org/10.1016/j.inffus.2014.01.006

Flanagan P (1995) Getting the paper out of the marketing & sales pipeline. Manag Rev 84:53–55

Foroudi P, Jin Z, Gupta S, Melewar T, Foroudi M (2016) Influence of innovation capability and customer experience on reputation and loyalty. J Bus Res 69:4882–4889. https://doi.org/10.1016/j.jbusres.2016.04.047

G2 Crowd (2017) Best marketing automation software in 2017|G2 Crowd. In: G2 Crowd. https://www.g2crowd.com/categories/marketing-automation. Accessed 1 Aug 2017

Gerrikagoitia J, Castander I, Rebón F, Alzua-Sorzabal A (2015) New trends of intelligent e-marketing based on web mining for e-shops. Proc Soc Behav Sci 175:75–83. https://doi.org/10.1016/j.sbspro.2015.01.1176

Goeldner C (1962) Automation in marketing. J Mark 26:53–56. https://doi.org/10.2307/1249632

Head G (1960) What does automation mean to the marketing man? J Mark 24:35–37. https://doi.org/10.2307/1248402

Järvinen J, Taiminen H (2016) Harnessing marketing automation for B2B content marketing. Ind Mark Manage 54:164–175. https://doi.org/10.1016/j.indmarman.2015.07.002

John P, Adlin T (2010) The persona lifecycle: keeping people in mind throughout product design. Morgan Kaufmann, MA/USA

Joshi S (2014) Customer experience management: an exploratory study on the parameters affecting customer experience for cellular mobile services of a telecom company. Proc Soc Behav Sci 133:392–399. https://doi.org/10.1016/j.sbspro.2014.04.206

Kang M, Sung M (2017) How symmetrical employee communication leads to employee engagement and positive employee communication behaviors. J Commun Manag 21:82–102. https://doi.org/10.1108/jcom-04-2016-0026

Kee A, Yazdanifard R (2015) The review of content marketing as a new trend in marketing practices. Int J Manag Acc Econ 2:1055–1064

Ki E, Ye L (2017) An assessment of progress in research on global public relations from 2001 to 2014. Publ Relat Rev 43:235–245. https://doi.org/10.1016/j.pubrev.2016.12.005

Kwan I, Fong J, Wong H (2005) An e-customer behavior model with online analytical mining for internet marketing planning. Decis Support Syst 41:189–204. https://doi.org/10.1016/j.dss.2004.11.012

Leeflang P, Verhoef P, Dahlström P, Freundt T (2014) Challenges and solutions for marketing in a digital era. Eur Manag J 32:1–12. https://doi.org/10.1016/j.emj.2013.12.001

Levy M (2015) Protecting customer data. Intern Audit 31–37

Levy Y, Ellis T (2006) A systems approach to conduct an effective literature review in support of information systems research. Inform Sci 0:181–212

Li S, Zheng Li J, He H, Ward P, Davies B (2011) WebDigital: a web-based hybrid intelligent knowledge automation system for developing digital marketing strategies. Expert Syst Appl 38:10606–10613. https://doi.org/10.1016/j.eswa.2011.02.128

Lichtenthal J, Eliaz S (2003) Internet integration in business marketing tactics. Ind Mark Manage 32:3–13. https://doi.org/10.1016/s0019-8501(01)00198-5

Longo D, Vilain P (2015) User scenarios through user interaction diagrams. Int J Softw Eng Knowl Eng 25:1771–1775. https://doi.org/10.1142/s0218194015710151

Luke K (2013) A little less automation, a little more personalization. J Financ Plann 26:20–22

Luxton S, Reid M, Mavondo F (2015) Integrated marketing communication capability and brand performance. J Advertis 44:37–46. https://doi.org/10.1080/00913367.2014.934938

Mac L, Shirley H (2015) The impact of internal marketing on organizational commitment: the mediating roles of consumer orientation and internal communication. Euro Asia J Manag 25:3–13

Martens D, Provost F, Clark J, Junqué de Fortuny E (2016) Mining massive fine-grained behavior data to improve predictive analytics. MIS Q 40:869–888. https://doi.org/10.25300/misq/2016/40.4.04

McAlister D (2006) The project management plan: improving team process and performance. Mark Educ Rev 16:97–103. https://doi.org/10.1080/10528008.2006.11488946

Moorthy J, Lahiri R, Biswas N, Sanyal D, Ranjan J, Nanath K, Ghosh P (2015) Big data: prospects and challenges. J Decis Mak 40:74–96

Peter M (2017) KMU-transformation: Als KMU die digitale transformation erfolgreich umsetzen. FHNW, Olten

Riecken D (2000) Introduction: personalized views of personalization. Commun ACM 43:26–28. https://doi.org/10.1145/345124.345133

Schäfer K, Kummer T (2013) Determining the performance of website-based relationship marketing. Expert Syst Appl 40:7571–7578. https://doi.org/10.1016/j.eswa.2013.07.051

Schoepf A (2017) Mehr Unternehmenserfolg mit Marketing Automation: Wie man automatisiert Neukunden generiert und bis zu 30% mehr verkauft. Books On Demand, Norderstedt

Schwaferts D (2016) Reif für den digitalen Wandel. UnternehmerZeitung 12

Singh K (2015) Creating your own qualitative research approach: selecting, integrating and operationalizing philosophy, methodology and methods. Vis J Bus Perspect 19:132–146. https://doi.org/10.1177/0972262915575657

SiriusDecisions (2017) Marketing automation platforms 2017. SiriusDecisions, Wilton CT/USA

Staatssekretariat für Wirtschaft (2017) Bericht über die zentralen Rahmenbedingungen für die digitale Wirtschaft. Staatssekretariat für Wirtschaft, Bern

Tiago M, Veríssimo J (2014) Digital marketing and social media: why bother? Bus Horiz 57:703–708. https://doi.org/10.1016/j.bushor.2014.07.002

Todor R (2016) Marketing automation. Bulletin of the Transilvania University of Brasov, *Series V: Economic sciences* 9(2):87–94

Wang E, Hu H, Hu P (2013) Examining the role of information technology in cultivating firms' dynamic marketing capabilities. Inf Manag 50:336–343. https://doi.org/10.1016/j.im.2013.04.007

Part III
Web 2.0 Revolution

FHNW Maturity Models for Cloud and Enterprise IT

Stella Gatziu Grivas, Marco Peter, Claudio Giovanoli and Kathrin Hubli

Abstract Existing cloud maturity models define the level of maturity according to the number of cloud solutions implemented or the duration of deployment. However, this does not include the motivation regarding why cloud solutions are used in an enterprise. Likewise, there has been no investigated into what changes the deployment of cloud services entails, how enterprise IT positions itself and whether the enterprise is able to bring about the necessary changes to support them. The combination of the FHNW Cloud and Enterprise IT Maturity Models provides information regarding why and how the cloud is used and how the enterprise IT is set up. As a result, the two maturity models support enterprises that already use the cloud on the way to making them even more efficient, to having their business IT aligned, and to contributing significantly to the successful digital transformation.

Keywords FHNW cloud maturity model · FHNW enterprise IT maturity model
Level of digital maturity · Digital transformation

1 Introduction

In the digitalization era in all of today's industries, existing business models are being disrupted more and more frequently and digital transformation dominates the management agenda. Information technologies such as cloud solutions, business

S. Gatziu Grivas (✉) · M. Peter · C. Giovanoli · K. Hubli
Institute for Information Systems, University of Applied Sciences and Arts Northwestern
Switzerland (FHNW), Von Roll-Strasse 10, 4600 Olten, Switzerland
e-mail: stella.gatziugrivas@fhnw.ch

M. Peter
e-mail: marco.peter@fhnw.ch

C. Giovanoli
e-mail: claudio.giovanoli@fhnw.ch

K. Hubli
e-mail: kathrin.hubli@fhnw.ch

© Springer International Publishing AG 2018
R. Dornberger (ed.), *Business Information Systems and Technology 4.0*,
Studies in Systems, Decision and Control 141,
https://doi.org/10.1007/978-3-319-74322-6_9

analytics and mobile technologies are enablers of the digital transformation (Gartner Inc. 2012; Châlons and Dufft 2015). Those who skillfully employ them stand out against competitors and thus create new business benefits (Blaschke et al. 2017). Enterprises that are able to integrate cloud approaches into their business and are also willing to allow changes are more efficient and thus more successful (Szymanski and Solis 2016). However, such a change obliges enterprises to seriously address their own situation and to rethink their own processes and business models (Blaschke et al. 2017). This also affects the IT organization, which is gradually becoming an internal IT service provider.

In the near future, IT departments will face new challenges: It has to become even more agile and efficient; it has to support business in the digital transformation and to reduce costs permanently. IT management including areas like business/IT alignment, IT governance and IT strategy will become crucial. New technologies will be used effectively and efficiently to support the development of new business models, optimize processes and most importantly to realize customer centricity.

This leads to new a role of the enterprise IT as a business partner and to a transformation from IT as a cost factor. The transformation of enterprise IT is of high complexity and does not include change in just one aspect, it means moving from a cost center to a business enabler, changing enterprise architecture to include the cloud, redesigning applications or selecting from shelf applications, moving from a centralized IT delivery to an IT service broker (Parekh 2014).

Consequently, for the evaluation of the cloud and enterprise IT there is a need for maturity models to support enterprises in analyzing their situation. This is necessary so that they can not only identify their current position, but can also see the discrepancy between their own enterprise and an optimal digital enterprise IT, or a cloud-enabled enterprise.

The FHNW Cloud Maturity Model and the FHNW Enterprise IT Maturity Model provide this. Both models are offered on the tool platform movecloud.ch[1], which is accessible only after registration. The online availability allows a maturity assessment that is immediately available to enterprises.

This book chapter is structured as follows: First, we will provide an overview of existing maturity models and contrast them with the FHNW maturity models. After that, we will explain both maturity models in detail in two separate sections. Then we will provide information on how to calculate the maturity level, how enterprises can use the maturity models and the evaluation of the models.

2 Analysis of Related Maturity Models

In the following section, we give a brief overview of existing maturity models both, for cloud and enterprise IT. This will result in a comparison of the models and in understanding their differences and will also help us to position our own maturity

---

[1] www.survey.movecloud.ch

Table 1 Comparison of the existing cloud maturity models

Model	Architecture & Infrastructure	Technology & Application	Information	Projects, Portfolios & Services	Operations	Organization	Governance	Cloud Operations	Provider Management	People	Enterprise Strategy	Security	XaaS Platforms	# Dimensions	# Maturity levels
Oracle	X		X	X	X	X	X							7	6
Gartner		X				X	X	X	X					5	5
Open Data Center Alliance	X	X	X	X	X	X	X				X	X	X	X	24	5
DAM Maturity Model	X		X		X		X			X		X		4	5

models in the research area. In our analysis, we focused on the specific areas—also known as dimensions—that each model considers for the calculation of the maturity level. More information on the dimensions will be given in Sect. 3.

The literature review of current cloud-related maturity models evidences that there are already some cloud-related maturity models in existence, such as the DAM Maturity Model (Davey et al. 2017) and maturity models from organizations such as Oracle (Mattoon et al. 2011), Gartner (Hilgendorf 2012) and Open Data Center Alliance (Colins et al. 2015). Table 1 provides a comparison of these cloud maturity models.

The literature review also evidences that there are already some enterprise IT related maturity models like the KPMG (KPMG AG 2016), the Adapt2Digital (Adapt2Digital 2015), the Altimeter Group (Szymanski and Solis 2016), the Arrk Group (Southward 2014), the Strategy and Transformation Consulting (STC 2016), Ernst and Young (EY 2016), the Hochschule St. Gallen-Crosswalk (Berghaus and Back 2016), and the IDC (IDC Research Inc. 2015). Table 2 provides a comparison of these enterprise IT maturity models.

Currently, no model has considered the connection between the use and integration of cloud solutions and the state of the digital transformation of an enterprise. However, cloud solutions can only make these changes efficient if they are optimally embedded in the enterprise, both strategically and operationally. By jointly looking at cloud maturity and transformation and by investigating which organizational changes are being favored or even required by the cloud, the FHNW Cloud Maturity Model builds a new bridge and closes this gap. In addition, the existing enterprise IT maturity models do not explicitly analyze the business/IT alignment or the applied IT service

Table 2 Comparison of the existing enterprise IT maturity models

Model	Strategy & Leadership	Culture	Governance	Data & Insights	Communication	Collaboration & Integration	Reporting & Monitoring	Customer Experience	Organization & Operations	Technology	Implementation	Risk & Cybersecurity	Finance & Legal	Product & Service	Employee	# Dimensions	# Maturity levels
KPMG	X	X					X	X	X	X					X	7	4
Adapt2 Digital	X		X	X	X	X	X	X		X	X					9	5
Altimeter Group	X			X	X			X	X	X	X					8	6
Arrk Group	X	X						X	X	X	X					5	-
STC	X	X	X	X	X	X	X	X	X	X					X	11	-
EY	X							X	X	X		X	X		X	7	-
HSG-Crosswalk	X	X				X		X	X	X				X		9	5
IDC	X	X	X					X	X	X					X	5	5

management within an organization. However, the FHNW Enterprise IT Maturity Model is able to offer new insights focusing on these aspects.

3   The FHNW Cloud Maturity Model

The FHNW Cloud Maturity Model includes four maturity levels with level 4 as the highest one. An overview and explanation of the maturity levels is given in Fig. 1. The maturity levels reflect an increasing maturity of the enterprise's cloud capability in the direction of digitalization. While cloud utilization is usually not yet being implemented at maturity level 1, the use of cloud services in maturity level 2 is actively addressed. A clear "cloud-first" strategy is being implemented in maturity level 3 while the final maturity level 4 focuses on optimizing the established cloud strategy.

The determination of the maturity level is carried out with the help of an online questionnaire. The questions are divided into six dimensions, which cover the following organizational, strategic and technical areas: cloud strategy, organization, knowledge and digital culture, governance, use of cloud services and architecture. The questions within a dimension are grouped according to different criteria. Table 3

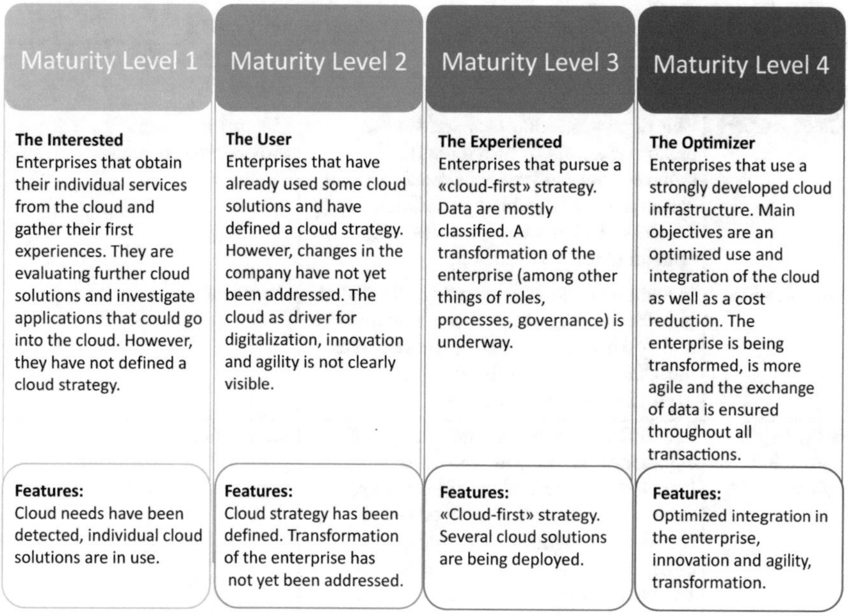

Maturity Level 1	Maturity Level 2	Maturity Level 3	Maturity Level 4
The Interested Enterprises that obtain their individual services from the cloud and gather their first experiences. They are evaluating further cloud solutions and investigate applications that could go into the cloud. However, they have not defined a cloud strategy.	**The User** Enterprises that have already used some cloud solutions and have defined a cloud strategy. However, changes in the company have not yet been addressed. The cloud as driver for digitalization, innovation and agility is not clearly visible.	**The Experienced** Enterprises that pursue a «cloud-first» strategy. Data are mostly classified. A transformation of the enterprise (among other things of roles, processes, governance) is underway.	**The Optimizer** Enterprises that use a strongly developed cloud infrastructure. Main objectives are an optimized use and integration of the cloud as well as a cost reduction. The enterprise is being transformed, is more agile and the exchange of data is ensured throughout all transactions.
Features: Cloud needs have been detected, individual cloud solutions are in use.	**Features:** Cloud strategy has been defined. Transformation of the enterprise has not yet been addressed.	**Features:** «Cloud-first» strategy. Several cloud solutions are being deployed.	**Features:** Optimized integration in the enterprise, innovation and agility, transformation.

Fig. 1 Maturity levels of the FHNW cloud maturity model

describes in detail the dimensions that are examined and matches them with the dimensions from Table 1.

Table 4 gives an example of how the dimensions, criteria, must-levels (ML), questions and maturity levels are linked. Each question for a respective criterion is associated with possible answers, different ones for level 1 to level 4. Level 4 represents the optimal state of the respective criteria in relation to the cloud maturity.

4 The FHNW Enterprise IT Maturity Model

As in the previous maturity model, the different maturity levels reflect the increasing maturity of the enterprise's digitalization (see Fig. 2). While at maturity level 1, digitalization is still largely rejected, its importance is already recognized at maturity level 2. At maturity level 3, there is a gradual implementation of digitalization activities or their preparation. At the final maturity level 4, digitalization is present in the daily routines, and the development of new business models is due to the high degree of maturity of the enterprise IT. Through all stages, the role of IT is transformed into an enabler and driver of digitalization in the enterprise.

In total, the following four maturity levels were defined for the digital maturity of the enterprise IT:

Table 3 Description of the dimensions of the FHNW cloud maturity model

Dimension	Description	Related dimension from literature analysis
Cloud strategy	This dimension addresses whether the enterprise knows its own (cloud) requirements and whether a cloud strategy has already been defined. In addition, the business/IT alignment is considered in relation to the cloud strategy	Enterprise strategy Provider management
Organization	In addition to defining a strategy for digital transformation, this dimension also involves seeing what new roles and practices were introduced to support the efficient deployment of cloud solutions	Organization
Knowledge and digital culture	In addition to organizational changes, a collaborative culture as well as existing knowledge are crucial for efficient cloud deployment. This dimension takes into account the existing knowledge in the enterprise and how the transfer of knowledge is executed. It also considers how to deal with innovation and new forms of working	Information people
Governance	This dimension considers whether policies and processes are established and whether committees that are appropriate for decisions were built. This dimension is devoted to the coordination of GRC provisions	Governance security
Use of cloud services	This dimension examines the way in which cloud services are deployed. More specifically, it examines whether their use is monitored, and the target achievement controlled, and whether appropriate adjustments are made to improve the use	Projects, portfolios and services Cloud operations XaaS platforms
Architecture	Concerning the architecture, it is considered which approaches are used for example for the orchestration or the network. Furthermore, the degree of virtualization plays an important role	Architecture and infrastructure Technology and application

For the FHNW enterprise maturity model, five strategic dimension (IT strategy, business/IT alignment, IT governance, IT leadership and organization as well as IT architecture) and four operational dimensions (IT service management, cloud computing, IT security and risk, IT controlling) were defined. These dimensions have in common that they are oriented towards the future, which primarily involves doing the right thing and thus pursuing an effective approach. Table 5 describes in

Table 4 Example of how the dimensions, criterion, questions und maturity levels are related

Dimensions	Criteria	ML	Question	Level 1	Level 2	Level 3	Level 4
Cloud strategy	Cloud need identifica-tion	2	Do you know your situation and your needs con-cerning cloud?	We partly know our situation and we are evaluating our cloud needs	We know our situation and our cloud needs	We know our situation and our cloud needs and we are adjusting our busines	We know our situation and our cloud needs in detail and we are optimising them
Cloud strategy	Cloud need identifica-tion	2	Have you defined cloud as a part of your strategy?	Cloud is not yet part of our strategy	Cloud is part of our strategy, but not com-pletely defined	Cloud is a com-pletely defined part of our strategy	Cloud is a com-pletely defined part of our strategy. We adjust and optimise it regularly

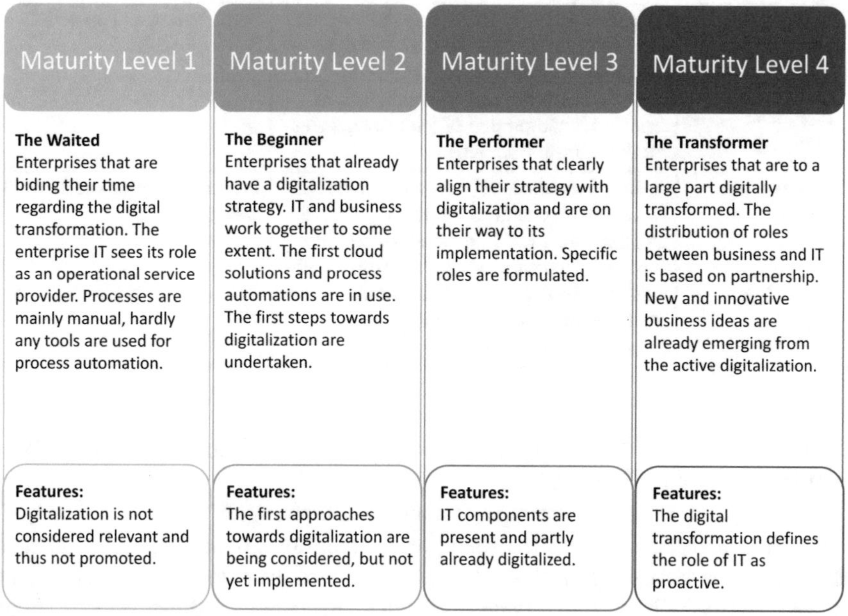

Maturity Level 1	Maturity Level 2	Maturity Level 3	Maturity Level 4
The Waited Enterprises that are biding their time regarding the digital transformation. The enterprise IT sees its role as an operational service provider. Processes are mainly manual, hardly any tools are used for process automation.	**The Beginner** Enterprises that already have a digitalization strategy. IT and business work together to some extent. The first cloud solutions and process automations are in use. The first steps towards digitalization are undertaken.	**The Performer** Enterprises that clearly align their strategy with digitalization and are on their way to its implementation. Specific roles are formulated.	**The Transformer** Enterprises that are to a large part digitally transformed. The distribution of roles between business and IT is based on partnership. New and innovative business ideas are already emerging from the active digitalization.
Features: Digitalization is not considered relevant and thus not promoted.	**Features:** The first approaches towards digitalization are being considered, but not yet implemented.	**Features:** IT components are present and partly already digitalized.	**Features:** The digital transformation defines the role of IT as proactive.

Fig. 2 Maturity levels of the FHNW enterprise IT maturity model

Table 5 Description of the strategic dimensions of the FHNW enterprise maturity model

Dimension	Description	Related dimension from literature analysis
IT strategy	In the ideal case, the IT strategy derives from the corporate strategy and is complemented by a digitalization strategy. The analysis of big data in its role as an innovative business driver is part of the strategy	Strategy and leadership
Business/IT alignment	Within the scope of the business/IT alignment, the cooperation between business and IT is illuminated, the customer being the focus of activities. Structural and social alignment are evaluated as important factors for successful digitalization	Customer experience Implementation communication
IT governance	The presence of a specific IT governance, within the framework of corporate governance, which is characterized by the use of methodologies such as ITIL, COBIT, etc., is regarded as fundamental to the digitalization maturity of IT. The accessibility of information via e-governance represents the optimized state	Governance
IT management/organization	This dimension, in addition to the new roles to be created for digitalization, which are partly accompanied by changes in the skillsets of employees and executives, is reflected in the support of the management for digitalization activities	Organization and operations Employee culture
IT architecture	A central dimension for digitalization is the IT architecture defined by the IT organization. In addition to architectural concepts, elements such as big data, mobile working, the use of platforms for the exchange of data and the degree of automation are analyzed	Technology collaboration and integration

detail the dimensions that are examined in the strategic dimensions and matches them with the dimensions from Table 2.

In addition to IT service management, cloud computing, IT security and risk, IT controlling also counts among the operational dimensions, which primarily increase efficiency. The increasing profitability and capacity of the enterprise are paramount in the described dimensions. Table 6 describes in detail the dimensions that are examined in the operational dimensions and matches them with the dimensions from Table 2.

Table 6 Description of the operational dimensions for enterprise IT maturity

Dimension	Description	Related dimension from literature analysis
IT service management	This dimension addresses the methodologies used in IT service management, as well as aspects such as incident management and possible automation options in sub-areas. The coordination of outsourcing partners and their coordination in the sense of a non-disruptive approach are highlighted	Governance Product and service
Cloud computing	The use of cloud technologies and their anchoring in the IT strategy up to a formulated complementary cloud strategy are the core issues in the field of cloud computing. The motivation to make use of the cloud is also being questioned	Technology data and insights
IT security and risk	In the context of modern forms of working, IT security and IT risk are measured at an elevated level. In particular, data security and the safety requirements for modern working methods such as mobile workstations and the necessary encryptions are investigated for the purpose of digitalization	Risk and cybersecurity
IT controlling	The measurability of the measures within the scope of digitalization also includes separate IT controlling which focuses on the degree of automation and also covers the real-time availability of important data and information for the controlling	Finance and legal reporting and monitoring

5 Calculating the Maturity Level of Enterprises

The assessment of the maturity levels of particular enterprises takes place on two levels. On the one hand, a separate maturity profile is created for each dimension and displayed graphically using a spider diagram. On the other hand, the company receives an overall assessment across all dimensions. The overall evaluation of the maturity level corresponds to the average of all maturity levels of the individual dimensions.

Two factors determine the maturity level within a dimension. First, the meaning of the corresponding responses is determined. A second step checks whether all so-called must-levels are fulfilled. For some questions, a certain maturity level is defined as a must-level. To achieve a certain maturity level, it is imperative that the companies that participate reach the must-level for the respective question. Thus, through the must-levels, a company that has more in-depth knowledge in complex areas, can

Dimension	Criteria	Question	Must-Level		Answer
dimension X	c1	q1	2		3
		q2	2		3
	c2	q3	3 ←	→	2
	c3	q4	2		3
	c4	q5	3		4
		q6	4		3

Mean value	3.0
Maturity level of dimension X	2

Fig. 3 Example evaluation per dimension with must-level

nevertheless fall back to a lower degree of maturity, if the mandatory must-levels are not fulfilled.

The following example illustrates this: The dimension X consists of four criteria (c1–c4) and comprises six questions in total (q1–q6). All the questions have different must-levels. The enterprise's responses resulted in a mean of 3.0. Nevertheless, the evaluation of the enterprise results in a maturity level of 2 for dimension X.

In this example, the self-assessment of question q3 resulted in a maturity level of 2. In order to achieve a maturity level of 3 in dimension X, the question q3 must result at least in a maturity level of 3. Since this must-level was not fulfilled, the enterprise will automatically fall back to a maturity level 2 in this example. An illustration of this evaluation example is shown in Fig. 3.

The allocation of the criteria is not specifically marked and therefore not apparent to the user. The determination of the levels for each criterion was developed based on the expertise of several specialists at the Competence Center for Cloud Computing. The various maturity levels were addressed and an investigation was conducted to identify the areas that are indispensable for each level.

6 How Enterprises Can Use the Maturity Models

The FHNW Cloud Maturity Model and the FHNW Enterprise IT Maturity Model are offered on the tool platform www.survey.movecloud.ch, which is accessible only after registration. After registration, participants are free to fill out the questionnaire whenever it suits them. Figure 4 gives an example of what the questionnaire looks like. The different dimensions are listed on the right. The questions are on the left. The participants choose picks the answer that fits best.

An evaluation example is given in Fig. 5. The different dimensions can be seen on the right. The dimension cloud strategy was selected for this example. The diagram is visible for the respective enterprise (red) and for the average of all participants (blue).

Fig. 4 Example online questionnaire

Letters A–E represent the different criteria. By means of integrated benchmarking, enterprises learn how they perform compared with other enterprises. Different factors can be used for the comparison. For example, enterprises can benchmark themselves within the same industry, region, or other statistical data. The stated results from the industry are both aggregated and anonymized.

Besides the spider diagrams for each dimension, users also receive an individual profile with a brief summary of their own situation in a second launch. Based on these profiles per dimension, specific recommendations are submitted. For the evaluation, the assigned maturity level is taken into account for each question and thus influences the generated recommendations. In this way, the enterprise learns in which areas there is a need to catch up in order to improve its own maturity level. However, the evaluation does not give guidelines on how to use the cloud or how to structure the enterprise IT.

EVALUATION: CLOUD MATURITY

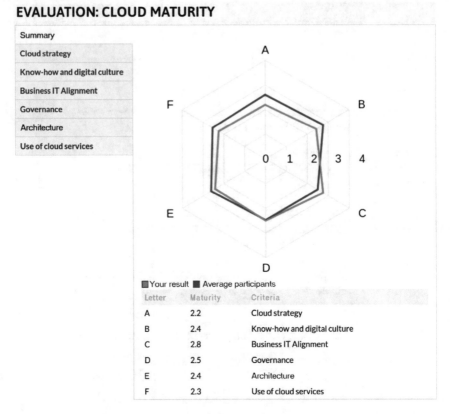

Fig. 5 Sample evaluation for a specific enterprise

The maturity models can also be used periodically to observe maturity over a longer period. Thus, the maturity model makes a crucial contribution in reviewing whether and how strategies and decisions affect the enterprises' own clouds and their corporate IT maturity. A repeated use of the models is certainly advisable.

Online availability gives companies multiple benefits all in one:

- The result of the maturity assessment is immediately available for the enterprises.
- The result can be exported as a PDF.
- After registration, the result can be saved and is therefore still accessible later. When repeating the assessment, for example, after important changes or new processes, the new result can be compared with the previous one.

7 Evaluation of the FHNW Maturity Models

The FHNW CxO Study was conducted on the basis of the FHNW Cloud Maturity Model and the FHNW Enterprise IT Maturity Model. This study was carried out for the first time in the first half of 2017, and will be repeated annually in order to analyze developments in the industry. The study design follows three specific objectives:

1. The state of transformation of IT drivers and inhibitors: Is transformation in Swiss enterprises already an issue, and how is it addressed? What are the drivers and what are the inhibitors?
2. The role and influence of cloud solutions: Is the cloud an enabler of digitalization as well as a catalyst for transformation?
3. Cloud maturity in Switzerland: Benchmarking with the maturity models: How fit are Swiss enterprises in terms of cloud deployment and corporate IT?

The study was designed as a qualitative survey based on interviews with IT executives and business representatives who are at the interface to IT. The interviews were used to evaluate the maturity models and, in particular, the dimensions and maturity levels. A summary of the results was published in a study report in September 2017.

8 Conclusion

Existing maturity models examine the maturity level of an enterprise with the focus on its digitalization, IT, or cloud capability. Until now, however, the maturity assessment has not been directed towards the application of cloud solutions and business/IT-alignment as well as IT service management. The two FHNW maturity models developed resolve this issue and provide a sophisticated maturity assessment for enterprises. The FHNW Cloud Maturity Model and the FHNW Enterprise Maturity Model offer their support as valuable tools for enterprises that use the cloud on their way to transformation. Focusing on the best possible integration of the cloud on a strategic as well as operational level, they lead the enterprise through an independent evaluation of its own cloud and its general corporate IT maturity level and thus its current situation. Based on predefined criteria, the maturity models show enterprises where there is a need to catch up and give recommendations—where appropriate. Thus, they are valuable tools that support enterprises in making important decisions.

The relevant recommendations and conclusions the participants are given after filling out the online questionnaire enable them to better understand their own situation and make appropriate decisions. Thus, the enterprises' attention is drawn to optimization potentials or concrete deficits, and they receive concrete hints to how these can be improved to increase maturity. In this way, the model helps enterprises to better integrate cloud solutions, thereby advancing the process of digital transformation.

However, the FHNW Maturity Models do not aspire to demonstrate how to use the cloud in an enterprise "correctly" or "wrongly", or how to set up enterprise IT appropriately. Recommendations are only made regarding how to reach the next level, and thus depend on the individual maturity level of the enterprise. Furthermore, the FHNW Maturity Models do not guarantee business success if the recommendations are implemented.

References

Adapt2Digital (2015) What stage of transformation is your business in? Adapt2Digital. http://www. adapt2digital.com/digital-maturity-assessment-1/. Accessed 25 July 2017

Berghaus S, Back A (2016) Gestaltungsbereiche der digitalen transformation von unternehmen: entwicklung eines reifegradmodells. Die Unternehmung 70:98–123. https://doi.org/10.5771/ 0042-059X-2016-2-98

Blaschke M, Cigaina M, Riss UV, Shoshan I (2017) Shaping the digital enterprise. Springer International Publishing, Cham

Châlons C, Dufft N (2015) Die Rolle der IT als enabler für Digitalisierung. In: Was treibt die Digitalisierung?: Warum an der Cloud kein Weg vorbeiführt. Springer Gabler, Wiesbaden

Colins A, Jung C, Regener I, Muench L-M, Maluf M, Estes M, Skipp R, Scott T, Dupley W (2015) Open Data Center Alliance Executive Overview—Cloud Maturity Model Rev. 3.0. Open Data Cent Alliance Inc. pp 1–22

Davey M, Regli T, Durga A, Elias M, Lipsey D, Dohr L (2017) DAM Maturity Model.http:// dammaturitymodel.org/. Accessed 22 July 2017

EY (2016) The EY digital readiness assessment (DRA). EY. https://digitalreadiness.ey.com/. Accessed 24 July 2017

Gartner Inc. (2012) Gartner says nexus of forces social, mobile cloud and information—Is the basis of the technology platform of the future. Gart. Inc. http://www.gartner.com/newsroom/id/ 2097215. Accessed 25 July 2017

Hilgendorf K (2012) Solution Path for public cloud adoption maturity plan. Gart. Inc. https://www. gartner.com/doc/2232218/solution-path-public-cloud-adoption. Accessed 22 July 2017

IDC Research Inc. (2015) IDC MaturityScape: digital transformation. IDC Res. Inc. http://www. idc.com/getdoc.jsp?containerId=254721. Accessed 24 July 2017

KPMG AG (2016) Digital readiness assessement. KPMG AG. https://atlas.kpmg.de/business-assessments/digital-readiness-assessment.html. Accessed 24 July 2017

Mattoon S, Hensle B, Baty J (2011) Cloud computing maturity model guiding success with cloud capabilities. Oracle 13

Parekh P (2014) IT Transformation: understanding environment and business goals. Retrieved September 3 2017 from http://blogs.cisco.com/datacenter/it-transformation-understanding-environment-and-business-goals

Solis B, Szymanski J (2014) Digital transformation how businesses are becoming increasingly digital. SlideShare. http://www.slideshare.net/Altimeter/slides-digital-transformation-with-brian-solis/10-Social_medias_progress_is_often. Accessed 22 July 2017

Southward I (2014) Digital transformation best practice thoughts. Arrk Gr. http://www.arrkgroup. com/thought-leadership/digital-transformation-best-practice-thoughts/. Accessed 22 July 2017

STC (2016) Strategy & transformation consulting—digital maturity assessment: bestimmung des digitalen reifegrades im unternehmen. Digit Transform Mag 1:16–19

Szymanski J, Solis B (2016) Six stages of digital transformation. ALTIMETER, A Prophet Co 55

Digital Transformation Management and Digital Business Development

Dino Schwaferts and Shama Baldi

Abstract In a study on the subject of digital transformation carried out by the University of Applied Sciences and Arts Northwestern Switzerland from April to May 2017, 82% of the companies stated that their driver for digital transformation was their wish to support their business processes. However, the majority of companies also stated that, in addition to the desired process support, they observed a greater impact on individual elements of the business model. The aim of this chapter is to explain these observations and their correlations, as well as to provide an approach regarding how digital transformation could be managed systematically against the background of its impact on the business model. Thereby, the terms "Digital Business Development" and "Business Stream Matrix" are used. Building on this approach of digital transformation management, the question is raised as to whether classical management concepts are still adequate to guide companies during and after digital transformation. Where there are management gaps, this chapter substantiates approaches to close these gaps. The term "digital leadership" is used. This chapter concludes with an approach to the "sociotechnical framework", outlining and substantiating a possible form of organization to which digital transformation tends.

Keywords Digital transformation management · Digital business development
Customer centricity · Digital society · Digital age · Business development teams
Risk-allocation · Business Stream Matrix · Meritocracy · Digital leadership
Sociotechnical framework · Ecosystem · Digital arena

1 Introduction

The chapter aims to describe the concept of digital transformation and its relevance to a dynamic digital society. Digital transformation is often misunderstood as the

D. Schwaferts (✉) · S. Baldi
Institute for Information Systems, University of Applied Sciences and Arts Northwestern
Switzerland, Riggenbachstrasse 16, 4600 Olten, Switzerland
e-mail: dino.schwaferts@fhnw.ch

© Springer International Publishing AG 2018 147
R. Dornberger (ed.), *Business Information Systems and Technology 4.0*,
Studies in Systems, Decision and Control 141,
https://doi.org/10.1007/978-3-319-74322-6_10

implementing of new technologies to improve or transform business operations or business functions. However, digital transformation is closely linked to business development in a systemic way that requires managing knowledge and skills as explained in Sects. 3 and 4. After we had gained insight into the concept and its relevance, and familiarize ourselves with the components of digital transformation, several approaches were developed to help manage digital transformation. The developed methods were verified with representatives of Swiss companies. The methodology used is based on a study conducted to identify barriers, drivers and the risks related to digital transformation and digital business development. Based on the study, we identify needs and the status quo of challenges and risks to digital transformation and digital business development. The identified needs and status quo helped to develop and verify approaches for managing digital transformation.

2   The Wrong Understanding of Digital Transformation

In the months from April to May 2017, the University of Applied Sciences and Arts Northwestern Switzerland conducted a study on digital transformation, in which 2,590 people from 1,854 companies participated.

In reply to the question regarding the motives for their digital transformation activities, 82% of the companies admitted to the desire of wishing to implement more efficient production and business processes. In addition to the possibilities of classical business informatics, they also rely on cloud solutions.

In reply to the question concerning the impact of digital transformation, 85% of the companies reported both the desired process improvements, and an impact on the business model or elements of the business model such as processes, customer relationships and products. Sixty nine percent of the companies see possible impacts on corporate culture, 62% on the market position and 50% on corporate culture.

However, adjustments or enhancements to the business model were never mentioned as the primary motive of digital transformation but only as a possible effect.

To see the difference between the objectives of classical business informatics and digital transformation, we take a brief look at the accommodation business: A hotel might consider using information technology to support the management of bookings, and the maintenance of rooms, to provide WLAN to the guests, to support billing and many other services. However, all these efforts are in vain if the customers of today's digital society book their stay via Airbnb. Airbnb is a great example of disruption in a traditional industry. It does not own any hotels or apartments. In other words, Airbnb does not have the operations of the affected traditional industry. Its strength lies in its access to and contact with the customer. Similar examples of digital transformation can be seen in other businesses, for example, fintech (Wadhwani et al. 2016) in the financial sector, Uber (Notelet 2015) in the transportation sector and the use of wearables (Coile 2000) in the health care sector. What we currently perceive is a transition into a digital society that affects nearly all aspects of our social life and all sectors of our economy.

Thus, the most exciting question is not how to optimize existing business operations or business functions using information technology, the cloud or digital services, but rather, how to identify, create and manage the needs of the coming digital society (the customers) and how to meet these needs in a competitive way. Adopting a customer-centric approach, wherein a company thoroughly focuses on identifying customer needs and managing positive customer experience is crucial for digital transformation. If a traditional company meets these upcoming needs, it will undergo a digital transformation, and will manage customer needs as a starting point for Digital Business Development.

3 Customer Needs for Digital Business Development

According to the previous section, customer centricity and the identification of coming needs are crucial for competitiveness in the digital society. Satisfying additional needs means developing additional products or services for these needs—in other words, developing additional business opportunities. Therefore, digital transformation is closely linked to business development.

The importance of customer centricity and the identification of upcoming needs are further enhanced by the great agility and fast development of new or adjusted needs. The following aspects make the alignment of value creation and customer needs particularly challenging:

1. *New customers are not known*: It is not enough to focus on the existing customers in order to develop additional business because digital transformation opens up possibilities to target new customers and new customer segments. Customers and customer segments that might be relevant for additional business can and usually will be unknown, as indicated in the previous section. What makes the identification of upcoming customer needs in the digital society difficult is the fact that we often do not know in which direction digital needs will develop. Therefore, it is a risk to limit the identification of customer needs to a few business areas such as marketing or top management. It seems to be more helpful to be open to all new ideas and to reflect to which extent other (all) employees can be involved in the identification of new customer needs.
2. *Use of communication channels*: There are a huge number of possible communication channels in the digital society, e.g. Facebook or Instagram. This is not enough, however. The company can possibly identify upcoming needs in live chats, on Twitter, Xing or specific blogs. However, the persons that are involved in the development of new business are not always familiar with the latest communication channels in the digital society or do not have sufficient time for their use. Not using a channel for the matching of customer requirements and desires or using an inappropriate or outdated channel may mean the loss of a business opportunity and pose a threat to the creation of meaningful and valuable success of digital transformation.

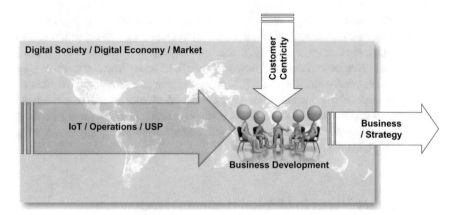

Fig. 1 Digital business development

3. *Early adoption*: New possibilities to meet customer needs can be very innovative and are based on very little experience. However, sufficient knowledge about these new digital possibilities is required to assess new ideas to meet these needs. In addition, new ideas may quite often become very complex when combining the early adoption of new digital possibilities with innovative approaches for customer relationships. For example, if someone plans to scan a room in three dimensions using a smartphone (to recommend or sell furniture), he or she could consider Google's ARCore (former Tango). However, he or she would then be one of the first to use this and would require the relevant knowledge to explore all possibilities.

Figure 1 schematizes Digital Business Development. In the digital future, customer centricity—identifying customer needs and meeting them with the appropriate business operations—is crucial for the development of business. Customer needs are the starting point for digital business development. However, as has already been pointed out, it is a risk to limit the identification of customer needs and digital business development to traditional areas like marketing or the top management level. A more team-based approach in the business strategy may be required.

4 The Significance of a Team-Based Approach

As mentioned in the previous section, knowing and managing customer needs is crucial for digital transformation. The identification of customer needs and business development will require a team-based approach—customer centricity should no longer be understood as an activity solely dedicated to the marketing department. Rather, it should become the task of all employees within an organization. Every employee can be the first source to initiate innovative ideas to bring in additional business. For instance, Starbucks, Southwest Airlines and Virgin America have involved their employees tin becoming part of the brand's marketing strategy (Parr 2012; Gallo 2013). This has helped them to better understand their customers' needs and to gain more transparency about their expectations. Employee engagement inspires customer confidence, which in turn drives sales. Thus, employees' engagement with customers builds stronger customer relationships. However, the involvement of employees in the identification of customer needs and in the development of a business has an impact on the expected employee engagement and the skills required:

- *Knowledge about marketing and business*
  Employees who have contact with potential customers need a knowledge of marketing and an understanding of how to design business models. In addition, they should be aware of the specific possibilities of their company regarding differentiation and uniqueness (see the three arrows in Fig. 1).
- *Commitment to the use of social media*
  Employees should obtain permission to use social media in their workspace and should have time to access it and reflect on it. The use of social media can help them to generate innovative ideas for additional business and to live customer centricity. Social media provides a means to foster employees' prospects as individuals with their own stories and possible needs. It serves to understand what makes customers happy and to derive factors that will lead to customer satisfaction.
- *Team-oriented culture*
  Employees should have the opportunity to share their thoughts. As pointed out before, the intensification of innovative ideas for additional business requires knowledge about possible customer needs in marketing, business development, new digital possibilities, and about the specific possibilities of their company regarding differentiation and uniqueness. Therefore, a team-oriented evaluation of ideas for digital business development needs to be supported and appreciated by the company culture.

5   Limitation of the Risk of Business Model Development

The ability to innovate remains the key source of creating and maintaining competitive advantage. However, innovation is not always associated with success; it may involve risk because of the uncertainty about the innovation itself and the costs that are incurred for the development of the additional business models. As defined by Al-Debei: "A business model is an abstract representation of an organization, be it conceptual, textual, and/or graphical, of all core interrelated architectural, co-operational, and financial arrangements designed and developed by an organization presently and in the future, as well as all core products and/or services the organization offers, or will offer, based on these arrangements that are needed to achieve its strategic goals and objectives" (Al-Debei et al. 2008).

Thus, we need to think about the possible risks before investing in business model development, e.g. what to do if the business idea is not successful. Certain aspects make this a greater challenge in the context of digital transformation:

- *Uncertainty regarding the development of the digital society*
  Obviously, business model innovations are always based on assumptions, be it customer behavior, competitive environment or governmental regulations (Miller 1994). However, because we do not know in which direction and at what pace the digital society will develop, it would be sensible to diversify efforts in different directions.
- *Cultural risk aversion*
  The digital economy is almost completely exposed to global competition. Therefore, the business model risk leads to a disadvantage for countries with a lower level of risk acceptance, such as Switzerland—while countries with a "do not be afraid to fail" mentality can benefit from global competition.

If companies from countries with a lower level of risk acceptance are to develop business models, the FHNW concept of Business Stream Management, which supports risk allocation and coverage, might be a solution (Schwaferts and Zhong 2016). The concept is based on the basic concept of the BCG matrix (García-Granero et al. 2015; Guță 2017). Based on the underlying ideas of the BCG matrix, and taking account of a less risk-taking management culture, the FHNW (University of Applied Sciences and Arts Northwestern Switzerland) developed the transformative Business Stream Matrix to help risk-averse managers to transform their business with transparent risk allocation and limitation. The risk limitation is supported by the overview, which allows the allocation of a positive cash flow pro rata with an expected success to the business development teams. Risk allocation means involvement of the business development teams in risk-acceptance. Engaged team members are likely to invest time in the development of a new idea, which can go beyond working time.

The matrix is classified according to two dimensions: Business Stream Environment and Competitiveness. Each dimension consists of three levels. In combination, they create a nine-cell-matrix as demonstrated in Fig. 2.

Derived from Al-Debei et al. (2008), a business model is a brief outline of arrangements needed to achieve strategic objectives. As a transformation is a development over time, we introduce the term "Business Stream" to highlight its development in an exponentially changing environment. The "Business Stream Environment" dimension represents the pace of change that a business idea is exposed to. If the idea of a business development team can be assigned to one pace of change, we represent it in the matrix as only one business stream. If the idea of a business development team has to be assigned to two or more different paces of change, it is a good idea to divide it up in two or more business streams and to represent and manage them separately. Otherwise, it would be more difficult to manage the Business Development Team.

The dimension "Competitiveness" refers to the estimated competitiveness of a product or service in a particular business stream. It can achieve unclear, clear or unique competitiveness. This classification supports the allocation of a positive cash flow to the business development teams.

Fig. 2 Business stream
matrix

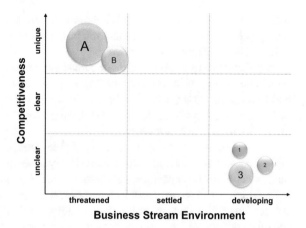

The innovations from the business development teams are highly likely to happen in a dynamically developing environment and are thus entered in the Business Stream Matrix on the right (see Circle 3 in Fig. 2). The Business Stream Matrix also makes it possible to add the existing business models, which will then be found on the left side or in the middle, if there are currently no disruptive dangers visible.

Business streams and business models can be entered as circles. With the size of a circle, management can visualize the cash flow that exists between business models and business streams, thus limiting the risk. The concept of cash flow here is similar to that in BCG Matrix (García-Granero et al. 2015; Guță 2017). The size of the circles can represent management importance in economic or strategic term. For example, if a new business stream has not yet generated sufficient cash surplus, but might require further investments, the size of the business stream is small. The example in Fig. 2 shows two existing business models, for which disruptive threats cannot be excluded, and a business stream that has not yet reached competitiveness and would still not be sufficient to support the company if business models 2 and 3 are disruptively weakened.

6   New Forms of Leadership and the Distribution of Risks

Prior to the Industrial Revolution, the average manufacturing company counted fewer than four employees. Goods were manufactured on a made-to-order basis in small workshops or at home (Koten 2013). Manufacturing was done mostly in homes using hand tools or very basic machines. With the development of steam engines, the Industrial Revolution started with an increase in volume and the introduction of mass production. Consequently, in the Industrial Age the predominant management style aims to push employees towards repetitive, measurable activities by instructions and controlling. Performance-related pay or Management by Objectives are examples of this. In addition, many other business elements such as working time regulations, workplace design, or personal evaluation instead of team evaluation all aim at clearly defined and measurable activities.

Fayol and Gray (1984) describes the management principles in the Industrial Age with "[t]o manage is to forecast and to plan, to organize, to command, to coordinate and to control." The question of how to integrate personal motivation into a team-oriented evaluation of ideas for business development was not part of the management principles for the Industrial Age.

Nowadays, we are seeing a reduction of the relevance of these management principles based on the tendencies to replace clearly described, repetitive and measurable human activities by machines, automation, artificial intelligence and physical systems, which goes hand in hand with a replacement of human management by technical control. "Anything that can be digitized or automated, will be" (Leonhard 2017). We see this tendency and the pace of change very clearly in Industry 4.0 concepts (West 2015).

On the other hand, we are seeing an increase in the importance of other success factors for human management. Another review of the Airbnb case shows that the business idea was created by two personally-motivated students. On the occasion of a major design conference in San Francisco in 2007, these students came up with the idea of renting out three airbeds on their living-room floor and of cooking their guests breakfast (Salter 2012). Since then, Airbnb has grown from one living room in San Francisco to over 200,000,000 guests in more than 65,000 cities and 191 countries (Airbnb n.d.; Guttentag 2015). At first, no one thought of it as a viable business model. The first attempts to make it work were almost impossible, and even today many Airbnb blocking regulations, such as emerging laws against the taxation of housing, threaten the business model (Guttentag 2015). We find analogue starting scenarios in many other examples of the digital economy, e.g. Google or Facebook. In all these examples, we have personally-motivated founders. However, the question that remains is how to integrate personal motivation into the previously described team-oriented digital business development.

To summarize and in the light of these facts, it is clear that we are indeed currently in the transition from management concepts for the Industrial age to management concepts for the Digital age. Traditional management concepts can become obsolete—while new management concepts are still being developed. The required management concepts will focus on those ingredients that cannot be digitized or automated, such as creativity, responsibility, innovation, passion and design. All these ingredients have one thing in common: They can neither be directly controlled nor prescribed by organizational regulations. A supervisor cannot give the command to be creative. The leading management can only work on the conditions that support these ingredients. We now briefly consider three conditions that have an indirect impact on the ingredients mentioned and therefore will have a significant leadership relevance in the digital economy:

1. *Flexibility*

   "It is not the strongest of the species that survives, nor the most intelligent that survives. It is the one that is most adaptable to change" (Darwin 1859). Flexibility or adaptability is a human characteristic—but can also be trained (Bailey 2014). A company can support this training, and a company can provide space and appreciation for flexibility. Google's famous "20% time policy" is a good example. "We encourage our employees, in addition to their regular projects, to spend 20% of their time working on what they think will most benefit Google" (D'Onfro 2015).

2. *Collaboration*

   Collaboration and personal networking are natural human instincts and can instinctively lead to collective intelligence. To develop collaboration, networking, and collective intelligence, the first and most important decision is not to disturb its natural emergence. That means abolishing all rules and practices that are contrary to the natural emergence of the desired activities. A further possibility to support the emergence of collective intelligence is to influence the team composition in order to balance the ingredients for human cognitive performance "stimulation", "synapses networking" and "valuation" (Schwaferts 2015).

3. *Meritocracy*

   Meritocracy describes a social system or an organization where the appreciation and impact or power of a person are not dependent on his or her position in the organization, but are a consequence of abilities, knowledge or intrinsic drive (Alexander 2016; Kim and Choi 2017). In digital transformation, the intrinsic drive is often the starting point. Entrepreneurial profit expectation is often an intrinsic motivation. Ability and knowledge might come with time. In addition to the above-mentioned examples (Airbnb, Google, Facebook), great examples for meritocracy can be found in the open source community.

7 Organizations Transform Towards "Sociotechnical Frameworks"

As discussed in the previous sections, we are currently in the transition from management concepts for the Industrial Age to management concepts for the Digital Age. It was also discussed that meritocracy and the intrinsic motivation of employees may also include the expectation of participation in the economic success of an innovation. Thus, the digital transformation has a visible influence on the management. Next, we want to list a selection of factors that work towards an organizational change:

- *Flexibility/Agility*

  We have already seen that the momentum is increasing in the digital future. For companies, this means that they have to react more flexibly and be more agile. Smaller structures have often a certain advantage here (see Fig. 3). In addition,

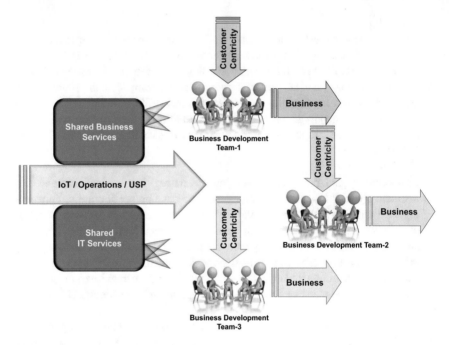

Fig. 3 Sociotechnical framework

short-term specific knowledge about an external person can be integrated faster than through a recruiting process.

- *Digitalization of Communication/Coordination*
  At the beginning of the Industrial Age, Coase (1937) described a balance that provides an explanation for the size of the company. Accordingly, the transaction costs of the procurement of performance influence the company size. If the transaction costs can be reduced by using digital technologies, the company can realize these cost savings by reducing its core size. In other words, if, with the help of IT, work coordination with an external partner works as smoothly as with an internal colleague, the involvement of external partners will promote flexibility and reduce the utilization risk (and thus costs).

- *Standardization*
  We can observe standardization efforts at almost all levels. Not only at the technical level, but also in business processes, for cloud services, etc. These standardization efforts support cross-company collaboration and thus smaller business units.

- *Risk Reduction*
  Higher risk components of overall corporate management can be outsourced to legally independent elements. This is also very easily transferable to digital business development. The committed employee, who is hoping for entrepreneurial success from the innovation, has a greater degree of willingness to bear the costs and expenses.

- *System Structures*

  Digitalization offers options for building system entry barriers. Apple demonstrates this with its eco-system (Nielson 2014) as clearly as the trend described by Fujitsu towards digital arenas (Tanaka 2016). The employee who is involved in digital business development and thinks entrepreneurially will probably also consider the option of whether, with her or his innovation, he or she has to rely on cooperation with his or her source company. If the source company has created systemic structures for system loyalty within its innovation project, it is more likely to benefit from employee commitment and the innovative strength.

This list is not conclusive, but it sensitizes us to the fact that the digital transformation is also associated with organizational transformation, and that this tendency will very likely lead to a reduction of the size of the company and to a development of ecosystems or digital arenas.

Transferred to the integration and management of employees or staff teams who drive digital business development with their entrepreneurial thinking, this means that when several small team islands with structural independence are created in a company, its innovative power is strengthened and risk and cost are reduced.

We are not yet able to document the manifestation of this trend for the digital future. Nevertheless, we introduce the concept of sociotechnical frameworks and sketch them in Fig. 3 as follows.

Digital transformation does not solely depend on digital technology but also on how individuals integrate themselves within an organization to transform their work. There is a need of interaction between work processes, technology and social systems involving people at workplaces so that the design and the implementation of customer needs are fulfilled in a holistic way. As mentioned in Sects. 5 and 6, in the "Business Development Teams", team members will hope to participate in a possible profit. Thus, we have entrepreneurial thinking. In entrepreneurial thinking, the team has the possibility to decide where to buy (obtain) required input. The inputs could be IT services, administrative support, operations support, etc. as mentioned on the left hand side of Fig. 3. With this holistic approach to digital transformation, the companies can convert to a sociotechnical framework.

8 Conclusion

Digital transformation is not solely about optimizing existing business operations or business functions with the help of information technology, by the use of the cloud, or by digital services, but is also closely linked to business development. It is about identifying, creating and managing the needs of the coming digital society (the customers). The alignment of value creation and customer needs require appropriate knowledge and skills because, without them, new customer segments cannot be identified, communication channels are not properly utilized and new possibilities are not well explored. To meet the challenges of aligning customer needs with business development, different approaches based on customer centricity were developed. One of the approaches emphasizes the significance of a team-based approach to digital transformation. Identified risk factors associated with innovation and digital business development are uncertainty regarding the development of the digital society and cultural risk aversion. To help manage the risks in digital business development, a Business Stream Matrix has been developed. The transparency of the Business Stream Matrix supports risk allocation. In addition, we identified the elements that could strengthen leadership relevance in the digital economy, namely, flexibility, collaboration and meritocracy.

The developed approaches were verified with representatives of Swiss companies in 40 digital maturity assessments and a study. The result of the maturity assessments suggests that the visualization in Fig. 1 (Digital Business Development) helps companies to achieve transparency in digital transformation. The results of the team-based approach are currently perceived as the recommendations with the highest priority level for risk management. The Business Stream Matrix is used in discussions and helps to find and manage widely diversified possibilities. However, the Business Stream Matrix remains on a theoretical surface that sets limits for this approach. The tendency to sociotechnical frameworks is currently of great interest to companies. It therefore stands to reason that research in this area has high priority.

References

Airbnb No Title. https://www.airbnb.com/about/about-us

Al-Debei MM, El-Haddadeh R, Avison D (2008) Defining the business model in the new world of digital business. In: AMCIS—Americas conference on information systems

Alexander D (2016) Meritocracy 2.0: A framework for decision-making. Open source

Bailey S (2014) 4 tips for being more flexible and adaptable. Bus J

Coase RH (1937) Nat Firm 4:386–405

Coile RC Jr (2000) The digital transformation of health care. Phys Exec 26:8–15

D'Onfro J (2015) The truth about Google's famous "20% time" policy. Bus. Insid

Darwin C (1859) Origin of species by Darwin, 1st edn. John Murray, London

Fayol H, Gray I (1984) General and industrial management. Institute of Electrical and Electronics Engineers

Gallo C (2013) How Southwest and Virgin America win by putting people before profit. Forbes

García-Granero A, Llopis Ó, Fernández-Mesa A, Alegre J (2015) Unraveling the link between managerial risk-taking and innovation: the mediating role of a risk-taking climate. J Bus Res 68:1094–1104

Guță AJ (2017) The analysis of strategic alternatives using BCG matrix in a company. In: 7th international multidisciplinary symposium

Guttentag D (2015) Airbnb: disruptive innovation and the rise of an informal tourism accommodation sector. Curr Issues Tour 18:1192–1217. https://doi.org/10.1080/13683500.2013.827159

Kim C-H, Choi Y-B (2017) How meritocracy is defined today?: contemporary aspects of meritocracy. Econ Sociol 10:112–121. http://dx.doi.org/10.14254/2071789X.2017/10-1/8

Koten J (2013) A revolution in the making. Wall Str J

Leonhard G (2017) Technology, humanity and the future of collaboration. https://www.zdruzenje-manager.si/assets/Uploads/Gerd-Leonhard.pdf

Miller KD (1994) Purdue e-Pubs diversification responses to environmental uncertainties

Nielson S (2014) Why Apple's ecosystem is its biggest competitive advantage. Mark, Realis
Notelet C (2015) Digital transformation of the economy: the case of Uber. Digit Bus, Strateg
Parr S (2012) Culture eats strategy for lunch. Fast Co
Salter J (2012) Airbnb: The story behind the $1.3bn room-letting website. Telegraph
Schwaferts D (2015) Komplexitätsbewältigung mithilfe emotionaler Bewertung. In: Hüsselmann C, Seidl J (eds) Multiprojektmanagement. Symposion Publishing
Schwaferts D, Zhong VJ (2016) Transformative business stream matrix. In: Tao M (ed) 2016 international conference on management science and engineering, 23rd annual conference proceedings. IEEE, Olten, pp 923–928
Tanaka T (2016) How does the evolution of digital technology impact on business and society?
Wadhwani R, Andrus G, Henry P, Kejriwal S (2016) Digital transformation in financial services
West DM (2015) What happens if robots take the jobs? The impact of emerging technologies on employment and public policy

Using Feedback Systems Thinking to Explore Theories of Digital Business for Medtech Companies

Michael von Kutzschenbach, Alexander Schmid and Lukas Schoenenberger

Abstract The rapid innovation of digital technologies poses a significant challenge to the healthcare sector. Digital technologies are transforming stakeholder relationships among established industry actors, including those of manufacturers, hospitals, and patients. To be ahead of competitors and to maintain profitability, medical device technology manufacturers (medtech companies) are urged to shift their business focus from product to customer excellence and thus invest in service offerings, focusing on the costs of alternative value delivery and patient outcomes. Such investments require a systemic and holistic understanding of how these changes in strategy affect the external and internal competitive environment. In this chapter, we propose the use of feedback systems thinking to explore the intended and unintended consequences of shifts in strategy, from sequential value chains to platform-oriented thinking. Taking the perspective of a medtech company in the value chain, we highlight challenges arising from hidden limits to growth that prevent the realization of intended achievements. Based on this, we develop hypotheses for the intended and unintended consequences of investing in digital service offerings. We conclude with a discussion of how systems thinking and modeling can support digital strategy development.

Keywords Digital transformation · Systems thinking · Systems archetypes
Business model · Theory of business · Digital service strategy · Platform
business · Healthcare · Medtech

M. von Kutzschenbach (✉)
Institute for Information Systems, University of Applied Sciences and Arts Northwestern Switzerland, Peter Merian-Strasse 86, 4052 Basel, Switzerland
e-mail: michael.vonkutzschenbach@fhnw.ch

A. Schmid
Institute of Information Systems, University of Liechtenstein, Fürst-Franz-Josef-Strasse, 9490 Vaduz, Liechtenstein
e-mail: alexander.schmid@uni.li

L. Schoenenberger
Department of Business, Health, and Social Work, Institute for Corporate Development, Bern University of Applied Sciences, Brückenstrasse 73, 3005 Bern, Switzerland
e-mail: lukas.schoenenberger@bfh.ch

© Springer International Publishing AG 2018
R. Dornberger (ed.), *Business Information Systems and Technology 4.0*,
Studies in Systems, Decision and Control 141,
https://doi.org/10.1007/978-3-319-74322-6_11

1 Introduction

Digital disruption is everywhere, and seems to be inevitable—a cliché that is regularly propagated in the media. Indeed, many industries have recently seen major shifts in competitive forces fueled by digitalization. The taxi industry and the hotel sector (viz., Uber and Airbnb) are often-discussed examples of the disruptive effects of digital technologies on businesses. However, while not all industries have experienced the same degree of disruption, the unreflective use and application of new digital technologies can have undesirable consequences. These may jeopardize or undermine the success of a business. Thus, in contrast to the uncritical optimism of technology evangelists and futurists, we agree with (Vermeulen 2017), who argues that some of the most common beliefs about the effect of digitalization on various industries have been "oversimplified, misunderstood, or misapplied".

Although the healthcare sector is slow to adopt digital technologies (Parente 2000; Wickramasinghe et al. 2005), changing stakeholder expectations and economic pressure are strong drivers of change. In addition, technology evangelists keep propagating digital technologies as saviors for these challenges. However, this stance neglects the dual role of digital technologies in organizational change. Such technologies can lock in processes as well as they can change them (Davies and Mitchell 1994; Easterbrook 2014; Peppard and Ward 2016). Thus, relying on dated approaches to the introduction of digital technologies for the transformation of organizations—for example, big bang implementations of new technology artifacts—is of no help for managing the messy, emergent process of a complex endeavor such as revising the business logics in an industry on the move (Weerakkody et al. 2011).

Feedback systems thinking offers an alternative approach. Systems thinking and modeling tools permit the analysis of potential consequences through the development of "micro-worlds" (Sterman 2001). These permit the operationalization of the "theory of business", our mental model of how a business works, to identify interactions between business models and interventions through the use of digital technologies (Drucker 1994; von Kutzschenbach and Brønn 2017). Such models capture essential causal relationships of planned transformation endeavors and enable the systematic evaluation of alternative approaches to technology implementation.

In this chapter, we propose the application of feedback systems thinking to the context of digital transformation. To this end, we have developed a case example which illustrates a medical device technology manufacturer (medtech company) that is planning to revise its strategic position. Consisting of more than a revision of its product portfolio, the change initiative is intended to transform the role of the company in the value chain. The development of this case example is based on interviews with managers of the company's leadership team. Applying a feedback systems approach, we develop hypotheses regarding the intended and unintended consequences of deploying innovative digital technologies for new service offerings in the healthcare industry.

To present the approach, we split the chapter into five sections. In the second section, we discuss the effect of digital technologies on the value chain. In the third section, we highlight the characteristics of a feedback systems perspective on digital transformation and its application to healthcare value chain. Based on the case example, the fourth section illustrates the potential of the feedback systems approach for the revision of theories related to the intended strategic change of the medtech business. This section further describes the intended and unintended implications of investments in digital service offerings for the role of a medtech company in the value chain. We conclude this chapter with a section reflecting on the strategic thinking required for a successful digital transformation of medtech companies and provide ideas for future research.

2 From Sequential Value Chains to Platform Businesses

Digital business transformation is a major challenge for all organizations, particularly in the healthcare industry (Bohlin et al. 2014). One reason for this is, as a long-serving chief physician of a major hospital puts it, the "…*remarkable IT-technophobia of healthcare organizations*". However, recent changes in consumer behavior and cost pressure on healthcare providers are forcing the industry to rethink the traditional value delivery model (van Amersfoort et al. 2014; McKinsey and Company 2017). While publicly-funded players such as many hospitals can afford a defensive position, this is a major issue for medtech companies that have recently experienced considerable pressure on profitability and growth, the latter substantially leveling off from 11% to 4% after 2008 (Belcredi et al. 2016). Thus, medtech companies in particular are being forced to rethink their business models for identifying emergent growth opportunities.

The traditional and currently dominant value chain in the healthcare industry, from the perspective of a medtech company, is sequential (i.e. a pipeline). In this simplified view, medtech companies deliver their goods to hospitals, who use them on the end user, the patients. The flow of goods is unidirectional. Hospitals serve as the connecting element between the device suppliers and the patients, who are the ultimate customers in the value chain (see Fig. 1). Due to system boundaries, the visibility of actions and interactions among players in the value chain is very limited for each entity. This is further restricted fragmented flows of information.

The emergence of innovative digital technologies such as mobile applications, cloud infrastructures, social networks, etc., promote the transformation of the value chain. Such technologies permit the revision of stakeholder relationships to create a more networked structure. Goods are no longer just physical, but can be digitally enhanced or are purely digital (e.g. information services, knowledge exchange, data provision etc.). Furthermore, information is becoming a strategic resource. Through such technologies and the accordingly increasing interconnectedness of actors, medtech companies and other stakeholders in the value chain may directly engage with patients, for example through digital services platforms. However, lack of standardization in processes, lack of (digital) competencies, and absent relational thinking in the value chain, obstruct the opportunity for medtech companies to shift

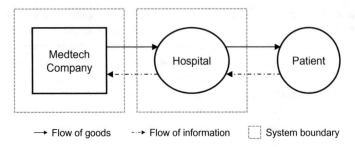

Flow of goods ···► Flow of information ▢ System boundary

Fig. 1 Healthcare as a sequential value chain

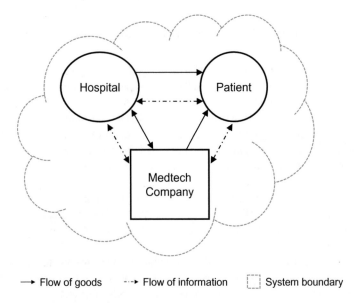

──► Flow of goods ···► Flow of information ▢ System boundary

Fig. 2 Healthcare as a platform business

into the role of a network-based mediator (Andal-Ancion et al. 2003) (see Fig. 2). Such a platform business strategy crucially relies on strategic alliances and partnerships with new and existing players in a tangle of complex relationships among participants of the healthcare sector—a so-called ecosystem.

The shift from pipeline firms (sequential value chains) to platform businesses with ecosystems redefines the boundaries of the established business environment (van Alstyne et al. 2016). A platform thus denotes two-sided or multi-sided markets where multiple stakeholders (e.g. companies or other industry actors) with cross-side network externalities can interact (Eisenmann et al. 2006; Rochet and Tirole 2003). All participants in a platform setup can incur costs and accumulate revenue. For platform businesses, the focus shifts from delivering one product to one type of buyer towards relationships and interactions—the exchange of value within the ecosystem. Therefore, the participants must deal with more pluralistic, complex and unstable environments. This is also because actors from outside the industry, e.g. from the information technology sector or the food industry, perceive growth possibilities in healthcare and are entering the market (Keys and Mainight 2010).

To take advantage of these trends, managers in medtech companies must rethink their "theory of business" (Drucker 1994). Those theories represent the organization's managerial understanding of "how things get done" (Osterwalder and Pigneur 2010), that is their mental model of how the business works. However, digital business transformation initiatives are cost intensive and uncertain. Increasing interconnectedness and the accelerating rate of change drive complexity, both within and across organizational boundaries (Billio et al. 2012; Kurzweil 2004). Thus, feedback systems thinking can help to create and analyze such a shared understanding of a business model's logic and its changes induced by digital business transformation initiatives.

3   A Feedback Systems Perspective on the Changing Healthcare Value Chain

The feedback perspective is a central element of systems thinking and modeling (Sterman 2000). It is best understood by comparing what is seen to be the "standard" approach to engaging with problems. The standard approach focuses on specific events that occur as a consequence of a problem-solving process. Decision makers compare the observations of these events with the desired situation. Where there is a discrepancy, a decision and the appropriate action is taken to minimize the deviation. The action and subsequent results conclude the decision-making process (see Fig. 3). If required, the next situation is then addressed. The main characteristics of this way of thinking are that it is linear and event-driven. There is usually no attempt to develop an operational explanation of the causes of the discrepancies. Such a way of thinking is particularly problematic in an environment where dynamic processes follow exponential developments, e.g. in networks (de Langhe et al. 2017; Senge 2006).

An alternative to the focus on events is the focus on feedback effects. Therein, the decision process recognizes the inherent dynamics of the situation, including the

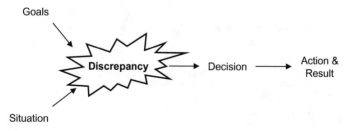

Fig. 3 Event-oriented world view (Adapted from Sterman 2000)

influence of actions taken in the past on the current situation. These, in turn, will influence future decisions. Additionally, this mode of thinking acknowledges the presence of other stakeholders and decision-makers who may have different goals and actions than the focal agent. A key aspect of this perspective is that it recognizes environmental impacts, meaning that one agent's actions to improve his/her situation affects another agent's ability to achieve his/her goals. This also includes the incorporation of unintended side effects on intended actions. Such effects influence the decision environment but often find no consideration in the mental decision models of individual actors. The presence of feedback structures implies that inputs are no longer independent of their outputs (Fowler 2003) (see Fig. 4).

Another characteristic of complex systems is the delay between cause and effect. Delays are inherent in organizational processes because responses to specific actions usually take a significant amount of time (Chen and MacMillan 1992; Larsen and Lomi 1999; Lomi et al. 2010). Consequently, the behavior of complex systems stands in contrast to open-loop, linear, sequential systems, and challenges traditional methods of analysis in which the independence, linearity, and strict exogeneity of influential factors are assumed.

Feedback systems thinking is a discipline that adopts a holistic perspective on complex organizational systems. The general approach is based on the system dynamics methodology that was initially developed by Jay W. Forrester at the Massachusetts Institute of Technology (MIT), USA, in the late 1950s. A systems thinking based analysis takes a step back from the level of single events and attempts to develop structural explanations of system behavior. Causal loop diagrams (CLDs) are a popular means of describing feedback loop systems. The core building blocks of CLDs are variables and causal relationships between them (see Fig. 5).

A causal loop diagram represents a feedback system. The loop blurs the distinction between the driver and the driven, between cause and effect, because, as time

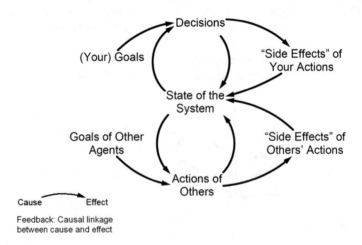

Fig. 4 Feedback-oriented world view (Adapted from Sterman 2000)

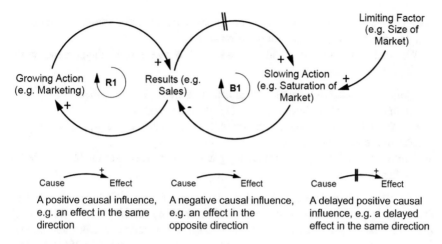

Fig. 5 Generic example of a causal loop diagram

progresses, each variable plays both roles. Taken together, the loops show overall systems behavior. Two types of feedback loops can be distinguished: positive or reinforcing (indicated by an "R" sign), negative or balancing (indicated by a "B" sign). Through identifying and mapping the causal linkages between variables, CLDs are representations of a modeler's understanding of systemic structures. The links shown by the arrows imply causal relationships between the different variables. This convention can express all causal relationships between and among variables of all kind. Variables cause change in other variables in two directions: "positive" (same direction, "+") or "negative" (opposite direction, "−"). For example, *Growing Action* in Fig. 5 is expected to increase (+) *Results*, and vice versa. Over time, however, as increasing *Results* is expected to increase (+) *Slowing Action,* which is expected to decrease (−) *Results* again. Time delays between causes and effects are marked with an "II" sign.

CLDs can be used to reveal managers' understanding of how a delineated system is designed and how it behaves. CLDs help to communicate and align the understanding of a person's or group's "theory of business". However, no model is ever complete, as each is an abstraction of reality, a reflection of the mental models of the people involved (Sterman 2002).

Recurring systemic structures responsible for generic patterns of behavior over time can be described as systems archetypes (Senge 2006; Wolstenholme 2003). Senge (2006, p. 93) describes them as follows: "If reinforcing and balancing feedback and delays are like the nouns and verbs of systems thinking, then the systems archetypes are analogous to basic sentences or simple stories that get retold again and again. ...As we learn how to recognize more and more of these kinds of archetypes, it becomes possible for us to see more and more places where there is leverage in facing difficult challenges, and to explain these opportunities to others." Thus, systems archetypes represent a thinking tool whose major purpose is to increase understanding of complex, dynamical systems or situations, and share insights about how the system in question works.

Wolstenholme (2003, p. 11) identified a set of four generic archetypes, each composed of two feedback loops:

- *"Underachievement*, where intended achievement fails to be realised;
- *Out of control*, where intended control fails to be realised;
- *Relative achievement*, where achievement is only gained at the expense of another [part of the system];
- *Relative control*, where control is only gained at the expense of others [stakeholders' benefits]."

We can distinguish problem archetypes from solution archetypes. A problem archetype is one whose net behavior is far from that intended by the people creating the system (see Fig. 5). The idea of a two-loop system archetype with problem behavior leads to the idea of a solution archetype to minimize side effects (Wolstenholme 2003). In this example, a significant problem for the system's long-term growth is market size. The CLD supports the identification of this limitation and shifts the attention from the intended consequences of "growing action" to "slowing action" and the according limitation. Managerial measures can be derived from this, e.g. the consideration of actions (introducing new "solution" feedback loops) to increase market size.

Digital transformation initiatives are an illustrative example of decision-making situations with multiple stakeholders and agents who are closely connected. A decision by one stakeholder will propagate through the system, affecting others, often with unknown consequences.

The simplified healthcare system we present in Fig. 1 has three major stakeholder groups—the medtech equipment supplier, the hospital and the patient. In their engagement with the value system, each stakeholder has dramatically different action sets and goals, many of which may conflict with the goals of the others. A successful transformation endeavor must take these issues into consideration. It can be speculated that a high percentage of unsuccessful transformation initiatives results from not recognizing the complexity of the change process (Flyvbjerg and Budzier 2011). From a feedback systems perspective, problem archetypes are pervasive in digital transformation endeavors.

4 Case Example—The Underachievement of DigitalMedTech

To illustrate the potential of system dynamics in the revision of theories of business, we have applied a feedback systems thinking approach to a medtech company. We derive our insights in part from interviews with executives in a major global medtech company which we call "DigitalMedTech" to preserve anonymity.

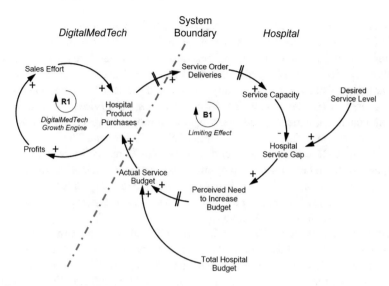

Fig. 6 DigitalMedTech problem archetype

We argue that with their current theory of business, DigitalMedTech faces an *underachievement* problem, which can be illustrated using a problem archetype. They focus primarily on the sale of physical goods. Increasing sales efforts result in higher hospital product purchases that, in turn, finance continued profit growth. We describe such a theory of business for DigitalMedTech in Fig. 6. Loop "R1" describes the growth engine. The hospital is a stakeholder in this process that has its own goals and limitations. A continuing increase in *"Hospital Product Purchases"* leads to an increase in the *"Service Capacity"* of the hospital. However, over time, the system will encounter a balancing process (Loop "B2") as the limit of the system is approached. This causes a delayed underachievement of the main objective over time (see Fig. 6).

A sequential value chain perspective strengthens the constraints imposed by existing system boundaries. The problem archetype in Fig. 6 includes a system boundary that "hides" the unintended consequences from the "view" of DigitalMedTech. Thus, upon realizing decreasing *"Hospital Product Purchases"*, managers can only invest in the enhancement of turnover, unaware that the amount of healthcare service spending depends on the *"Hospital Service Gap"*. In this model, the *"Hospital Service Gap"* results from the discrepancy between *"Desired Service Level"* and actual *"Service Capacity"*. The level of discrepancy induces the pressure for change. This translates into specific action to improve the situation, i.e. closing the gap between the actual and the desired states. Thus, an increase in hospital *"Service Capacity"* leads to a decrease in the *"Hospital Service Gap"* which causes a decrease in *"Hospital Product Purchases"* over time due to a limited *"Actual Service Budget"*. This ultimately reduces service purchases which adversely affects the growth model of DigitalMedTech.

A possible way out of this underachievement for DigitalMedTech is to break out of the isolated role in the value chain and to proactively manage its stakeholder relationships, introducing one or more "solution" feedback loops. Strategic partnerships could help DigitalMedTech to gain direct access to patients. Digital technologies, furthermore, could improve customer retention. Thus, value-based healthcare services (Foley et al. 2014), including digital services offerings such as patient education, patient engagement, operating room efficiency, rehab follow up, and outcome measurements, are a potential field of investment. The "solution" is to position DigitalMedTech to better understand customer needs (for hospitals as well as patients) and to help hospitals to optimize their process efficiencies for delivering better patient outcomes. Being able to extend its offerings by implementing digital services as described above permits DigitalMedTech to evolve its position to a more central role in the healthcare system.

Driving technology-enabled services in the healthcare system allows DigitalMedTech to offer added value to its strategic partners through capturing and comparing a variety of data from different companies (hospitals) as well as patients. This requires process standardization in order to enable comparability and digitalization for both DigitalMedTech and hospitals. The standardization of the processes, especially in the operating room (OR), enables benchmarking and thus further optimization which in turn will positively drive DigitalMedTech's sales. This has the dual effect of saving costs and improving the service capacity of hospitals.

In the next section, we describe what such a revised theory of business for DigitalMedTech could look like developing a solution archetype.

5 Digital Services Platform Business as Solution

On its way to extending the focus from product towards customer excellence, DigitalMedTech started an initiative investing in a digital patient platform and services. Based on Wolstenholme (2003) understanding of the solution archetype concept and on information gathered from interviews with representatives from DigitalMedTech's leadership team, we extended the "DigitalMedTech problem archetype" model (see Fig. 6) to a solution archetype. Thus, the mapping of the "solution links" results in two additional loops that enable the partnering hospitals to run their processes more efficiently, "R2: Efficiency Improvement Loop", and to improve patient engagement and outcomes, "R3: Quality Improvement Loop" (see Fig. 7).

"R2: Efficiency Improvement Loop": A higher investment in digital service increases the "*Degree of Process Standardization*" of healthcare processes which improves the "*Service Process Efficiency*", in particular OR efficiency. Thus, effective use of OR time is imperative for cost-efficient operations and close attention should be paid to practices that affect the efficiency of the OR. Standardizing processes and improving information increase transparency. Improving on-time starts and turnover time increases hospital production. Furthermore, reducing the cancellation rates helps to reduce unnecessary costs to OR use, and saves money which

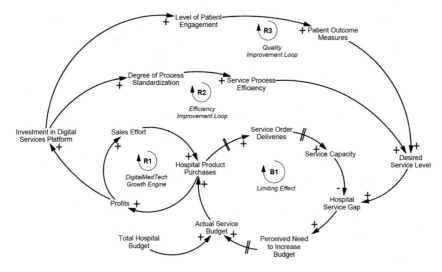

Fig. 7 DigitalMedTech solution archetype

would otherwise be spent on unnecessary setups, instrument sterilization, and supplies. Reducing cancellation rates and delays frees up availability in the OR schedule as well as maximizing OR use. Thus, improving *"Service Process Efficiency"* contributes to increase the *"Desired Service Level"*.

"R3: Quality Improvement Loop": Today 45% of consumers search for health information on social media (van Amersfoort et al. 2014). Social media and mobile platforms are becoming increasingly important channels for consumers and provide a means to measure patient satisfaction to improve the quality of healthcare delivery (e.g. Porter 2009; Porter et al. 2016). Higher investments in digital service offerings lead to a higher *"Level of Patient Engagement"*. Being able to leverage social media extensively to engage with consumers will help, firstly, to provide consumers with timely and relevant information, and secondly to improve feedback about patient satisfaction and service quality, thus improving the level of *"Patient Outcome Measures"*. Improving the level of *"Patient Outcome Measures"* will increase the *"Desired Service Level"* of the hospital.

"Investment in Digital Services Platform" is a key element of a digital strategy for DigitalMedTech. At the same time, it will face unanticipated and potentially undesirable consequences. Providing services based on emerging digital technologies often fosters a lock-in effect that is difficult to disengage from. For DigitalMedTech this might be preferable, however, for hospitals it can impose constraints that could be seen as undesirable.

Such effects can be illustrated with balancing feedback loops added to the model. They can feed back to DigitalMedTech's growth engine (see Fig. 8). The two negative feedback loops (Loop "B2" and "B3") are driven by an *"Investment in Digital Services Platform"*. This makes the hospital increasingly dependent on the knowledge and

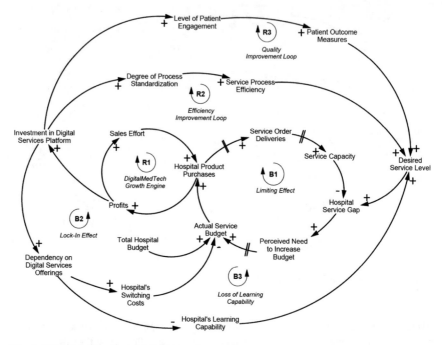

Fig. 8 Potential side effects of digital service strategy

technology of DigitalMedTech. First, loop "B2" shows the costs of being technically bound by increasing the "*Hospital's Switching Costs*" to another platform provider. This affects the hospital's budget, the "*Actual Service Budget*" because switching from established technology and connected services can be costly, and sometimes nearly impossible without starting from scratch. Second, loop "B3" is related to the limitations on the "*Hospital's Learning Capability*" by being bound to a single digital services supplier. Being locked-in makes it difficult for the hospital to invest in its own learning capability and to facilitate organizational learning, in particular double-loop learning (Argyris 1977). This can affect the "*Desired Service Level*" provided by the hospital. The "*Hospital Service Gap*" is the key driver in the hospital's budgeting process so this is an important relationship.

Feedback systems thinking enables the explication of theories of business (von Kutzschenbach and Brønn 2017). This makes them easier to evaluate and adapt. The tendency to focus on the desired outcome, and neglect or ignore the undesirable ones, is hazardous. Feedback systems thinking contributes to overcoming this effect by building models that can be explained, criticized, and modified by relevant stakeholders. An explicit model of the "theory of business" encourages active discussions of the critical assumptions underlying the current business model. In order to revise the model, the next steps will involve quantifying the variables and testing the assumptions in a simulation model. These steps allow the creation of a computer-based micro-world and a validation of the model.

6   Conclusion

Managers and organizations hesitate to engage in feedback systems thinking for problem solving. This is mainly because decision-makers operate under different and conflicting performance indicators in separate departments, teams, and functions. This fragmentation of the larger organization into smaller silo organizations in traditional value chains masks critical links and vital interdependencies amongst different actors with adversarial effects on the external and internal organizational environment.

The pace in today's organizations is so unrelenting that there is little room for managers to pause, reflect, and learn. Consequently, most organizations realize only small fractions of their potential (Forrester 1994). This is particularly striking in the context of digital transformation endeavors as consequences of such initiatives are nearly impossible to assess *ex ante*. Applying feedback systems thinking helps decision-makers to operationalize and communicate models for their "theory of business". Such an approach enables a better identification of interactions between the model and the digital intervention.

The DigitalMedTech case provides an illustrative example of industry players facing underachievement. Until now, they have barely been able to reap the potential benefits of emerging digital technologies. However, growing their business requires a shift from transactional, sequential value chains to a platform business. They would thereby aspire to act as trusted advisors to other healthcare providers such as hospitals and patients. A digital service platform using a systematic measurement of outcomes that mattered to patients would allow DigitalMedTech to provide value-based services for improving the healthcare delivery service with benchmarking and learning on a condition-by-condition basis. Driving such a digital transformation is a complex endeavor. We thus propose to use feedback systems thinking for the revision of theories of business.

We provide an extended causal loop model that can be used to develop a simulation model. This can enhance learning, provide deeper understanding and insights, and reveal inconsistencies and "blind spots" in policies and digital strategies. However, it is clear that the feedback systems thinking approach in managing digital transformation endeavors is only at its beginning. Further research is needed to connect feedback systems thinking and modeling to digital transformation, and demonstrate its potential benefits as a complementary tool for strategy development.

References

Andal-Ancion A, Cartwright PA, Yip GS (2003) The digital transformation of traditional businesses. MIT Sloan Manag Rev 44(4):34–41
Argyris C (1977) Double loop learning in organizations. https://hbr.org/1977/09/double-loop-learning-in-organizations. Accessed 2 Aug 2017

Belcredi A, Danger T, Rosenberg B, Gerecke G, van Duijnhoven H, Eichelberger M (2016) Medtech companies need to transform while times are still good. https://www.bcgperspectives. com/content/articles/medical-devices-technology-transformation-medtech-companies-need-transform-while-times-still-good/. Accessed 8 Aug 2017

Billio M, Getmansky M, Lo AW, Pelizzon L (2012) Econometric measures of connectedness and systemic risk in the finance and insurance sectors. J Financ Econ 104(3):535–559. https://doi. org/10.1016/j.jfineco.2011.12.010

Bohlin N, Kaltenbach T, Kharbanda V, Herzig S (2014) Succeeding with digital health: winning offerings and digital transformation. http://www.adlittle.com/downloads/tx_adlreports/ADL_ 2016_Succeeding_With_Digital_Health.pdf. Accessed 2 Apr 2017

Chen M-J, MacMillan IC (1992) Nonresponse and delayed response to competitive moves: the roles of competitor dependence and action irreversibility. Acad Manag J 35(3):539–570. https://doi. org/10.2307/256486

Davies L, Mitchell G (1994) The dual nature of the impact of IT on organizational transformations. In: Baskerville R, Smithson S, Ngwenyama O, DeGross JI (eds) Transforming organizations with information technology, vol 49, pp 243–261

de Langhe B, Puntoni S, Larrick R (2017) Linear thinking in a nonlinear world. Harvard Bus Rev 95(3):130–139

Drucker PF (1994) The theory of the business. https://hbr.org/1994/09/the-theory-of-the-business. Accessed 3 Aug 2017

Easterbrook S (2014) From computational thinking to systems thinking: a conceptual toolkit for sustainability computing. In: Proceedings of the 2nd international conference ICT for sustainability 2014

Eisenmann T, Parker G, van Alytsne MW (2006) Strategies for two-sided markets. Harvard Bus Rev 84(10):92–101

Flyvbjerg B, Budzier A (2011) Why your IT project may be riskier than you think. Harvard Bus Rev 89(9):23–25

Foley C, Kronimus A, Schenk M, Bielesch F (2014) The 2013 Medtech value creators report: finding sustainable value in a changing market. Accessed 2 Aug 2017

Forrester JW (1994) System dynamics, systems thinking, and soft OR. Syst Dyn Rev 10(2–3):245–256. https://doi.org/10.1002/sdr.4260100211

Fowler A (2003) Systems modelling, simulation, and the dynamics of strategy. J Bus Res 56(2):135–144. https://doi.org/10.1016/S0148-2963(01)00286-7

Keys T, Mainight TW (2010) Who is looking after you?: Blurring industry boundaries in health & wellness. https://www.slideshare.net/tskeys/who-is-looking-after-you-blurr. Accessed 2 Apr 2017

Kurzweil R (2004) The law of accelerating returns. In: Teuscher C (ed) Alan turing: life and legacy of a great thinker. Springer, Berlin, New York, pp 381–416

Larsen ER, Lomi A (1999) Resetting the clock: a feedback approach to the dynamics of organisational inertia, survival and change. J Oper Res Soc 50(4):406–421. https://doi.org/10.2307/ 3010461

Lomi A, Larsen ER, Wezel FC (2010) Getting there: exploring the role of expectations and preproduction delays in processes of organizational founding. Org Sci 21(1):132–149. https://doi.org/ 10.1287/orsc.1090.0437

McKinsey & Company (2017) Den digitalen Patienten verstehen. https://www.nzz.ch/marktplaetze/ sponsored-content-serie-digitalisierung-fuer-mckinsey/sponsored-content-serie-digitalisierung-fuer-mckinsey-8-den-digitalen-patienten-verstehen-ld.146613. Accessed 5 Feb 2017

Osterwalder A, Pigneur Y (2010) Business model generation: a handbook for visionaries, game changers, and challengers. Wiley, New Jersey, US

Parente ST (2000) Beyond the hype: a taxonomy of e-health business models. Health Aff 19(6):89–102. https://doi.org/10.1377/hlthaff.19.6.89

Peppard J, Ward J (2016) The strategic management of information systems, building a digital strategy, 4th edn. Wiley, New Jersey, US

Porter ME (2009) A strategy for health care reform—toward a value-based system. New Eng J Med 361(2):109–112. https://doi.org/10.1056/NEJMp0904131

Porter ME, Larsson S, Lee TH (2016) Standardizing patient outcomes measurement. New Eng J Med 374(6):504–506. https://doi.org/10.1056/NEJMp1511701

Rochet J-C, Tirole J (2003) Platform competition in two-sided markets. J Eur Econ Assoc 1(4):990–1029. https://doi.org/10.1162/154247603322493212

Senge PM (2006) The fifth discipline: the art & practice of the learning organization, Rev edn. Broadway Books, New York

Sterman JD (2000) Business dynamics: systems thinking and modeling for a complex world. McGraw-Hill Publishing Company, Boston, Mass

Sterman JD (2001) System dynamics modeling: tools for learning in a complex world. Calif Manag Rev 43(4):8–25

Sterman JD (2002) All models are wrong: reflections on becoming a systems scientist. Syst Dynam Rev 18(4):501–531. https://doi.org/10.1002/sdr.261

van Alstyne MW, Parker GG, Choudary SP (2016) Pipelines, platforms, and the new rules of strategy. Harvard Bus Rev 94(4):54–62

van Amersfoort M, Stegwee R, Jansen P, Buvat J (2014) Taking the digital pulse: why healthcare providers need an urgent digital check-up. https://www.de.capgemini.com/resource-file-access/resource/pdf/healthcare-provider-digital.pdf. Accessed 2 Apr 2017

Vermeulen F (2017) What so many strategists get wrong about digital disruption. https://hbr.org/2017/01/what-so-many-strategists-get-wrong-about-digital-disruption. Accessed 2 Apr 2017

von Kutzschenbach M, Brønn C (2017) Education for managing digital transformation: a feedback systems approach. J Syst Cybernet Informat JSCI 15(2):14–19

Weerakkody V, Janssen M, Dwivedi YK (2011) Transformational change and business process reengineering (BPR): lessons from the British and Dutch public sector. Govern Informat Quarter 28(3):320–328. https://doi.org/10.1016/j.giq.2010.07.010

Wickramasinghe NS, Fadlalla AMA, Geisler E, Schaffer JL (2005) A framework for assessing e-health preparedness. Int J Electron Healthcare 1(3):316–334. https://doi.org/10.1504/IJEH.2005.006478

Wolstenholme EF (2003) Towards the definition and use of a core set of archetypal structures in system dynamics. Syst Dynam Rev 19(1):7–26. https://doi.org/10.1002/sdr.259

Ontology-Based Metamodeling

Knut Hinkelmann, Emanuele Laurenzi, Andreas Martin
and Barbara Thönssen

Abstract Decision makers use models to understand and analyze a situation, to compare alternatives and to find solutions. Additionally, there are systems that support decision makers through data analysis, calculation or simulation. Typically, modeling languages for humans and machine are different from each other. While humans prefer graphical or textual models, machine-interpretable models have to be represented in a formal language. This chapter describes an approach to modeling that is both cognitively adequate for humans and processable by machines. In addition, the approach supports the creation and adaptation of domain-specific modeling languages. A metamodel which is represented as a formal ontology determines the semantics of the modeling language. To create a graphical modeling language, a graphical notation can be added for each class of the ontology. Every time a new modeling element is created during modeling, an instance for the corresponding class is created in the ontology. Thus, models for humans and machines are based on the same internal representation.

Keywords Modeling · Ontologies · Metamodel · Enterprise modeling
Domain-specific modeling language

1 Introduction

Decision makers use models to understand and analyze a situation, to compare alternatives and to find solutions. Business process models, for example, enable the identification of potential improvements and the communication of process variants with stakeholders. Enterprise models serve as a baseline for changing the enterprise. Engineers use models as blueprints for planning and construction.

K. Hinkelmann (✉) · E. Laurenzi · A. Martin · B. Thönssen
School of Business, FHNW University of Applied Sciences and Arts Northwestern Switzerland,
Riggenbachstrasse 16, 4600 Olten, Switzerland
e-mail: knut.hinkelmann@fhnw.ch

© Springer International Publishing AG 2018
R. Dornberger (ed.), *Business Information Systems and Technology 4.0*,
Studies in Systems, Decision and Control 141,
https://doi.org/10.1007/978-3-319-74322-6_12

Models describe and represent the relevant aspects of a domain in a defined language. There are many different kinds of modeling languages: graphical models, conceptual models, mathematical models, logical models. Even textual descriptions can serve as models. The choice of the modeling language depends on what the models is used for and who is using the model.

General-purpose modeling languages such as UML have the advantage that they can be used to represent any kind of information. However, they have the disadvantage that they do not guide people in modeling. People have difficulty conceptualizing the domain and different people might conceptualize the domain in different ways.

Domain-specific modeling languages, on the other hand, consist of modeling elements which have a pre-defined meaning that a domain expert can understand. Business process modeling languages, for example, are specialized for modeling the process flow using elements such as tasks and events and relationships to represent the order of the task execution. Domain-specific modeling languages reduce the degree of freedom for the modelers and thus support the understanding and reuse of models by different people.

Models are typically designed for a specific purpose. There are a huge variety of domain-specific modeling languages and modeling tools. Business Process Model and Notation (BPMN) has been designed to provide a standard visualization mechanism for business processes, which are defined in an execution optimized business process language (OMG 2011). BPMN engines allow the deployment and execution of business processes. Besides, process models can also serve the purposes of process optimization, governance, risk analysis and compliance management. Process models designed for execution, however, are often not compatible with models serving these purposes, although they have an overlapping set of modeling elements. Furthermore, there are typically specific tools for the various purposes, each with its own modeling language. Consequently, processes have to be re-modeled several times.

For a comprehensive view it would be beneficial to have modeling languages that serve several company-relevant issues such as decision-making, automation and compliance. Different applications can share or exchange models, or parts of them, and thus avoid re-modeling. As a prerequisite, the semantics of the modeling language has to be clearly defined.

Humans use graphical models for communication and to identify potential for improvement. Figure 1 shows an example of a business process model in BPMN (OMG 2011). Humans can recognize that there is a deadlock for a known customer in the customer lane and they can propose parallelization to check the formal and financial consequences, because there is no data dependency between these tasks.

In software engineering, graphical models of UML are widespread, e.g. using class diagrams for conceptual modeling. ArchiMate (The Open Group 2016) is a modeling language for enterprise architecture.

While graphical and textual representations are well understood by humans, they are not adequate for machine interpretation. Formal models are required to be interpreted by software systems must be formal models. Business process improvement can be supported in this with software tools, with which models can be checked

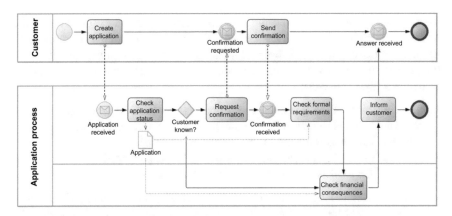

Fig. 1 Graphical business process model for human interpretation

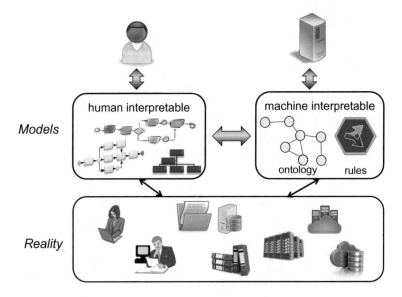

Fig. 2 Typical modeling approach: separate models for human and machine interpretation

for consistency, KPIs can be measured and simulations can be performed. A typical approach is to have separate models for humans and machines (see Fig. 2):

- Graphical notations which can easily be understood by humans are provided.
- Formal models such as databases, mathematical models and program code are used for machine interpretation.

The focus on machine-interpretable knowledge is known as knowledge engineering (KE) and is distinguished from knowledge management (KM), which focuses on human-interpretable knowledge (Karagiannis and Woitsch 2010). However, the models used by humans and machines are not strictly separated. Humans create the models and then use machines for analysis. The results are then again presented to humans for interpretation. However, if humans and machines use different models, it is hard to maintain consistency of the models. The use of both graphical and formal models has two challenges:

1. The semantics of the graphical and formal models must be identical. One way to achieve this is to define the semantics of both models in a formal language, which is known as semantic lifting (Azzini et al. 2013).
2. If part of the reality is represented in both graphical and formal models, a change in any of the models must be mirrored in the others.

Thus, a modeling language that can be interpreted by both humans and machines would be advantageous. This modeling language needs to be formal enough to be interpreted by a machine and must have a graphical presentation layer, which facilitates interpretation and manipulation by humans.

Figure 3 depicts the basic elements of the ontology-based modeling approach, which satisfies these requirements. Entities of reality are represented in graphical models suited for humans and at the same time in a machine-interpretable formal model (i.e. in an ontology). Both models are deeply intertwined. The ontology provides a formal semantics of the modeling language (Hrgovcic et al. 2013; Kappel et al. 2006) such that related models are interpretable by machine.

In our ontology-based metamodeling approach, we use ontologies to define the semantics of the modeling language in a formal model, which can be interpreted by a machine. A model engineer can represent the domain knowledge as an

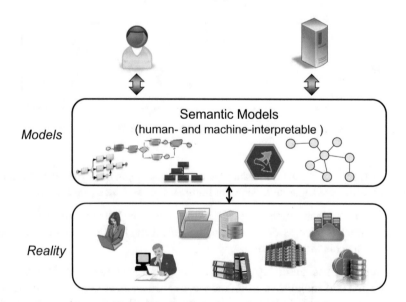

Fig. 3 Our proposal: integrated models for human and machine interpretation

ontology with classes, relations and rules. To create a graphical modeling language, the classes can be extended by graphical notations, which can be used by the modeler to create models. Thus, the ontology represents the domain knowledge, which at the same time is the metamodel of a domain-specific modeling language. By making a graphical model, the modeler creates instances of ontology classes. Thus, models are formal with clear semantics. As a result, the ontology-based metamodeling approach achieves two goals: (1) the definition of domain-specific modeling languages with an unambiguous formal semantics, for which (2) the models can be interpreted both by humans and machines.

2　State of the Art

In the following section, we detail research on modeling languages, its formalization for machine interpretation, and approaches to combine formal representation of models with models which are cognitively adequate for humans.

2.1　Modeling Languages

According to Karagiannis and Kühn (2002) a modeling method consists of a modeling language and a modeling procedure, as well as modeling mechanisms and algorithms. In the following, each of these three components is explained in the context of enterprise engineering.

Metamodels are the basis for modeling tools and for the development of the modeling languages. They provide the syntax of a modeling language. A metamodel contains the class hierarchy and the properties representing the modeling elements as well as the relations between them (Jonkers et al. 2003). This corresponds to the so-called abstract syntax. The specification of the graphical notation for each modeling element and relation corresponds to the concrete syntax. The latter should be cognitively adequate to ensure the users' understanding of models that are built from it. The domain-specific conceptualization addresses this aspect by providing modeling elements that are tailored to a given domain. Fill and Karagiannis (2016) analyzed the conceptualization of modeling methods: They use the ADOxx metamodeling platform[1] to investigate how to realize four selected functionalities of enterprise information systems to support user interaction, process-based optimization, interfaces to other systems, and complex analyses.

---

[1] ADOxx is a commercial product and trademark of BOC AG.

2.2 Machine Interpretability

To gain its full potential, the purpose of modeling must go beyond transparency and communication, which is what graphical models provide humans with. Models should also be used for automation, and operations such as decision making, analysis, adaptation and evaluation.

For automation purposes, model knowledge should be machine-interpretable or at least machine-readable. In business process automation, for instance, process models determine the workflow executed by the workflow engine. For decision-making purposes, it is common practice to work with models, for example, as represented by the Decision Model and Notation (OMG 2016).

In keeping with (Hinkelmann et al. 2016a), we distinguish between machine-interpretable models and machine-readable models by claiming that the former are represented in a format on which reasoning can be performed. Hence, machine-interpretable models can turn passive data storage into an active device. A machine-interpretable format can be expressed in logic-based languages such as ontologies. Different kinds of reasoning can be applied, depending on the expressivity of the ontology language. Ontologies expressed in the Resource Description Framework Schema (RDFS) (W3C 2014), for example, can be combined with semantic rules to draw new insights from the already existing knowledge base (KB).

2.3 Combining Human with Machine Interpretability

In the field of information systems, the human interpretability of modeling refers to metamodels, whereas machine interpretability mainly refers to the formal semantic aspects of models, i.e. ontologies (Hinkelmann et al. 2016a). Höfferer (2007) discusses the relationship between metamodels and ontologies by emphasizing that metamodels and ontologies are different but complementary concepts. Ontologies basically furnish both modeling language constructs and their instances with formal semantics (Dietz 2006; Kramler et al. 2006; Kappel et al. 2006). Metamodels on the other hand, mainly provide the syntax and graphical representation for those modeling language constructs. Aßmann, Zschaler and Wagner (2006) assume that ontologies in the Semantic Web and models in model-driven engineering (MDE) were developed in isolation and investigate the role of ontologies, models, and meta-models to bridge the gap between the two communities.

2.4 Semantic Lifting

Semantic lifting is defined as "…the process of associating content items with suitable semantic objects as metadata to turn 'unstructured' content items into semantic

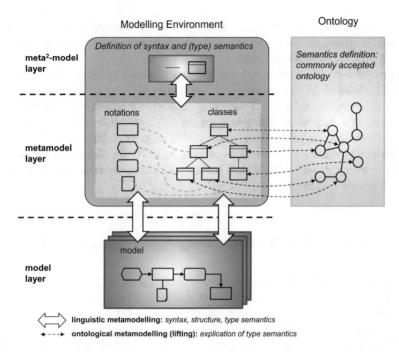

Fig. 4 Metamodels for human-interpretable and machine-interpretable models (Höfferer 2007)

knowledge resources" (Azzini et al. 2013). This approach requires the relationship between the human-interpretable and the machine-interpretable modeling languages to be defined (Hrgovcic et al. 2013).

The metamodels for the human-interpretable graphical representations and the machine-interpretable metamodels, e.g. represented in an ontology, are strictly separated. To align them, formal and non-formal (meta)models are mapped by transformation.

Figure 4 shows the conceptual architecture for semantic lifting. Different model types in the enterprise architecture are created which correspond to different metamodels. These primarily define syntactical aspects as well as certain semantic aspects of model elements. The ontologies define the machine-interpretable semantics of the modeling concepts. In the literature, semantic lifting is also known as semantic annotation (Liao et al. 2015; Fill et al. 2013).

In these approaches, the ontologies are independent from the concepts of the human-interpretable, graphical languages. The ontology comprises class definitions which represent the formal semantics of modeling elements. Furthermore, it includes class definitions which serve to annotate models and model elements. The basis for interoperability is provided by linking model elements of the models and metamodels with ontology concepts.

This approach has been described in and used, for example, in the European research projects LearnPAd (De Angelis et al. 2016) and CloudSocket (Hinkelmann et al. 2016b, c; Woitsch et al. 2016).

The drawback of this approach lies in the consistency of the semantics between (meta)models and their representation in ontologies. Keeping them separate tends to cause incompatible semantics. This mainly occurs if the project stakeholders do not agree among themselves on a common understanding of important terms beforehand, or if little attention is paid when changes occur, i.e. poor maintenance.

Having provided a brief overview of related work, we claim that human- and machine-interpretable models should become an integrated model in order to realize the full potential of modeling.

3 Conceptual Solution of Ontology-Based Metamodeling

In order to avoid the inconsistency problem between (graphically represented) models and ontologies, a semantic metamodeling approach is proposed which merges the abstract syntax of metamodels with the semantics defined in the ontology. This means that the ontology is used to specify both the semantics and the abstract syntax.

The ontology is extended by a specification of the graphical notation. The difference to the transformation approach is that the semantics is expressed only once for both human-interpretable and machine-interpretable models. The ontology-based modeling can be regarded as a variant of the MOF metamodeling framework (OMG 2014) where UML is replaced by an ontology language as a metamodeling language.

In the ontology-based metamodeling approach, the ontology itself is also the metamodel for the graphical modeling environment. Only the graphical notation for each concept is defined separately from the semantic description (see Fig. 5). A mapping is defined between concept definition and graphical definition (Nikles and Brander 2009).

The semantics is in the ontology, which consists of classes, attributes, relations and constraints. The model layer of Fig. 5 is an instantiation of both semantics and related notations that resides in the metamodel layer. Thus, the model in the bottom layer benefits from both, a semantics that is machine-interpretable and a graphical notation that makes it human-interpretable.

In addition, the ontology-based metamodeling approach fosters the adaptation of a modeling language to fit a specific domain. In order to have a common understanding of the term "adaptation" we refer to the work of Laurenzi et al. (2017), where the following operations were performed in the metamodel layer using existing modeling languages:

- Identification of needed and unneeded concepts
- Specialization/generalization of concepts
- Restrictions on attribute values
- Injection of constraints among concepts

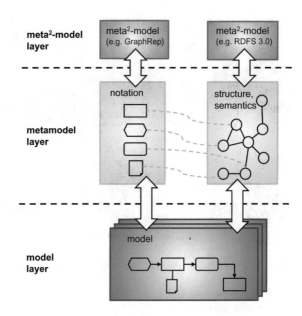

Fig. 5 Ontology-based metamodeling

Specializing/generalizing a concept can refer to both classes and relations. For example, the class "task" in BPMN generalizes the classes "user task", "manual task", "service task" and "business rule task". These can then be further specialized. For example, "user task" can be specialized into "send electronic document", "send package", etc. An example for domain restrictions for attribute values is described by Hinkelmann et al. (2016a), where an attribute that expresses the functionality of a cloud service can only have values from the APQC Process Classification Framework (APQC 2014). Injection of constraints refers to the additional relations that can occur among modeling elements. These restrict the way concepts can be connected in the models.

In the literature, these operations refer to the actions that typically take place at the design time of Domain-Specific Modeling Languages (DSMLs) (Fowler 2011, Frank 2010; Gray et al. 2008; Mernik et al. 2005; van Deursen et al. 2000). DSMLs shift the complexity of modeling from the model layer (M) to the metamodel layer (MM) (see Fig. 6). In the context of our ontology-based metamodeling, ontology experts work in the metamodel layer to make the modeling easier for the language user.

In the model layer (M), users make use of the constructs developed in the meta-model layer (MM) to create models. In Fig. 7 we provide an explanatory example, already used by Emmenegger et al. (2016). We assume, for example, that the modeling element "C1" reflects the class "Lane" of BPMN (OMG 2011), while "C2" reflects the class "Role" of the Organizational Model. By adding a relation "r" between the two modeling elements, we allow one or more instances of "Lane" to be connected with one or more instances of "Role". This enables the language user

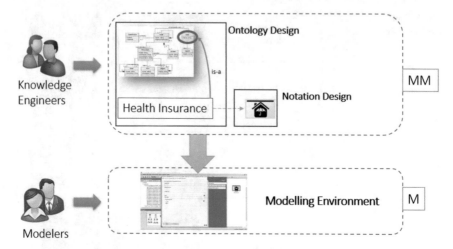

Fig. 6 Domain-specific conceptual modeling with an ontology-based metamodeling approach

Fig. 7 Two-tier approach. Adapted from (Laurenzi et al. 2017)

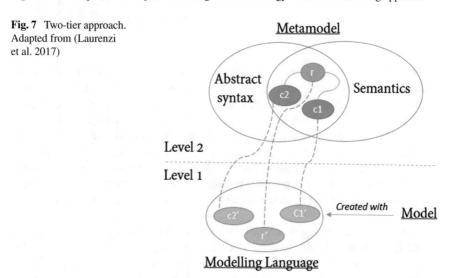

to refer a specific lane to a particular role so that, for example, the role can be further specified in an organizational model.

The degree of freedom in modeling in the model layer (also known as level 1) depends on the level of specificity of the modeling language (also called degree of semantics by Frank (2010)) inserted in the metamodel layer (also known as level 2). The higher the level of specificity is, the more domain-specific is the modeling language. The UML class diagram, for example, provides a general-purpose meta-model with a low specificity level. Hence, the language user can create and connect any classes without restrictions, i.e. the user has a high degree of freedom in the model layer. Such freedom may create ambiguous models, leading to the wrong interpretation of models or modeling the same reality differently, which can make the models questionable. Even worse, a low level of specificity can lead to nonsense models, inconsistency and wrong models. This is a critical issue, because, for example, in the context of Model-Driven Development, it can lead to

(a) the creation of error-prone or unintended software, and
(b) quality issues such as incomplete requirements or poor implementation.

The BPMN metamodel already provides a higher level of specificity than the UML class diagram. It contains

(a) a taxonomy of concepts (e.g. user task specifies task),
(b) attributes that might differ from concept to concept (e.g. task versus event), and
(c) constraints among concepts following the BPMN guidelines that allow the modeling of structured processes.

For example, a start event initiates the process, an end event ends the process and one or more activities should occur between the two events.

Due to their higher level of specificity, DSMLs promise to enable domain experts to handle the designing and editing of models in a meaningful and less error-prone way, and therefore support the production of high quality models (Kelly and Tolvanen 2008). Moreover, allowing the domain experts to deal directly with familiar language constructs makes the language easy to learn and improves its applicability (Hudak and Paul 1996). DMSLs offer the benefit of a high level of understanding of models among domain experts, fostering not only productivity in design time, but also the optimization phase, where pain points are rapidly identified and actions can be taken accordingly.

One of the main challenges of DSMLs is to inject the metamodel with the appropriate level of specificity. This challenge relates to the design of a DSML, which can be supported by a machine if the language is grounded with a formal semantics, i.e. an ontology-based metamodel.

4   Implementation

In this section, we present the implementation of the ontology-based metamodeling approach as it was developed in the European research project CloudSocket (Woitsch et al. 2016).

4.1   The BPaaS Ontology

The metamodel for the service selection, allocation and deployment is represented in the BPaaS Ontology. The ontology in Fig. 8 conceptualizes functional and non-functional specifications of a cloud service.

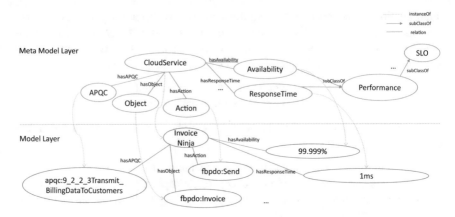

Fig. 8 Part of the BPaaS ontology developed in the European research project CloudSocket

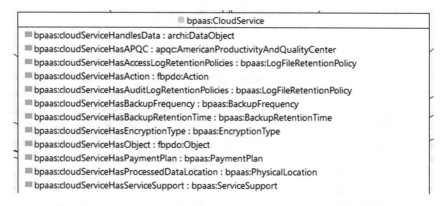

Fig. 9 The cloud service concept implemented in the BPaaS ontology

Figure 9 shows some of the attributes of the cloud service concept. The functional aspects specify functionalities of a cloud service. They relate to the hierarchy of the APQC Process Classification Framework (APQC 2014), using the relation "cloudServiceHasAPQC". In addition, cloud services functionalities are specified through actions and objects from a predefined taxonomy with the relations "cloudServiceHasAction" and "cloudServiceHasObject". This corresponds to the convention of BPMN to name activities using a verb (i.e. action) and a noun (i.e. object), e.g. "Send Invoice".

The model layer in Fig. 8 depicts the instances of the concepts in the metamodel layer. For example, APQC category, action and object are used to specify the functionalities of the cloud service "InvoiceNinja", i.e. "9.2.2.3 Transmit Billing Data to Customer", "Send" and "Invoice", respectively.

Furthermore, from the BPaaS Ontology a cloud service can also be specified through non-functional aspects, for example, availability and response time (see

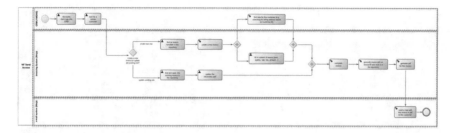

Fig. 10 Extension of BPMN 2.0 implemented in ADOxx

Fig. 8). These two refer to the performance category, which in turn is reported in the Service Level Objectives (SLOs) that is listed in the Cloud Service Level Agreement Standardisation Guidelines (C-SIG 2014). Figure 8 shows the conceptualization of the "availability" and "response time" as subclasses of the "performance" concept, which in turn is a subclass of the "SLO" concept. The bubbles containing values "99.999%" and "1 ms" are instances of the classes "Availability" and "Response-Time", respectively.

The graphical notation for the modeling elements in the model layer are added after the ontology design that takes place in the metamodel layer. In the following, we describe two different types of graphical representations that were implemented for the same ontology-based metamodel.

4.2 Model-Based Representation Implemented in the Metamodeling Tool ADOxx

The modeling language BPMN 2.0 was extended with the Service Description Model to specify both functional and non-functional aspects of a cloud service. The language extension was implemented in ADOxx. As Fig. 10 shows, each lane can be annotated with a cloud service description element, which is specified in the Service Description Model. Specifications occur in the notebook. Figure 11 shows the functional specifications, while Fig. 12 shows the non-functional specifications. The functional specifications in the notebook reflect the relations to APQC, Action and Object as defined in the BPaaS Ontology. Their values represent the instances in the model layer of Fig. 8. As soon as the notebook is saved, an ontology instance is created.

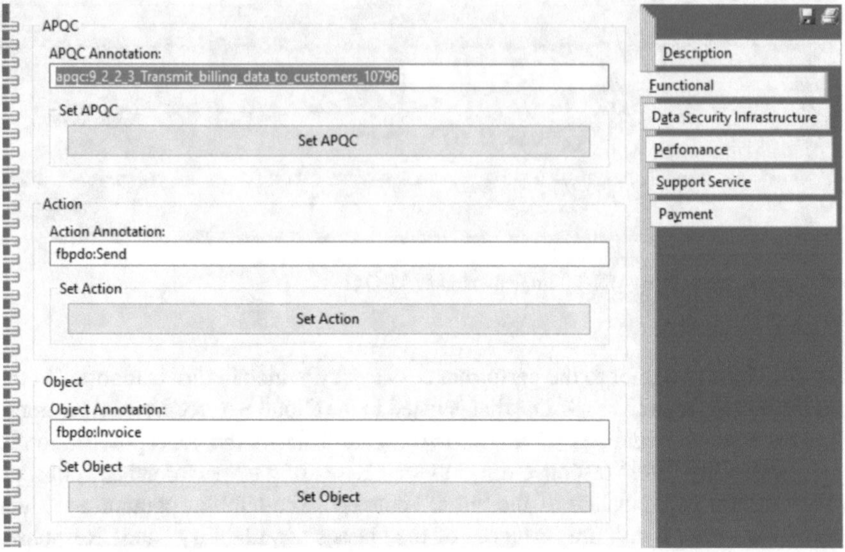

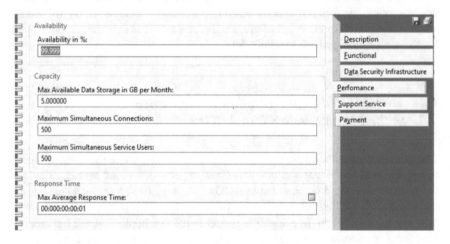

Fig. 11 Non-functional aspects of a cloud service implemented in ADOxx

Fig. 12 Non-functional aspects of a cloud service implemented in ADOxx

4.3 Web-Based Representation

Figure 13 shows the representation of the cloud service (functional and non-functional) specifications implemented in a web-based format. Instead of creating a graphical process model, the functional and non-functional requirements are specified in a web form. The web form is a human-oriented modeling language, which is used as an alternative to the graphical representation of the process using BPMN.

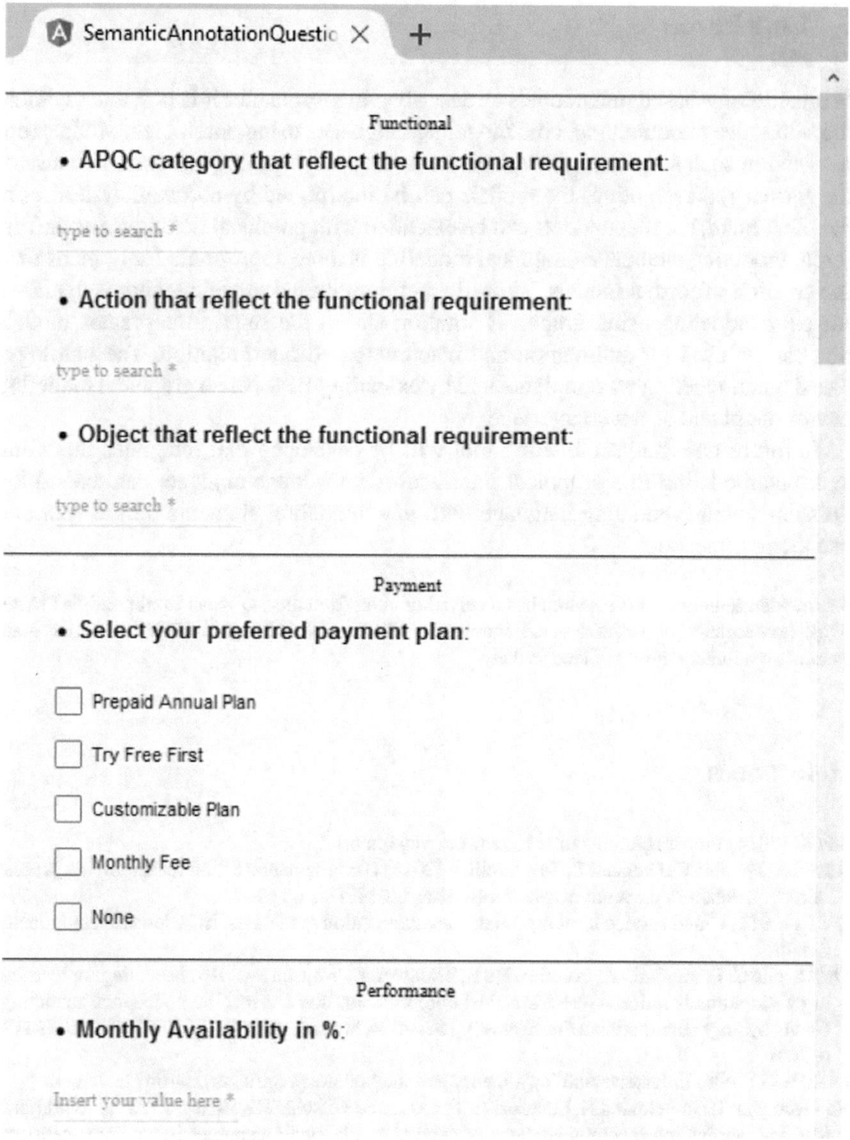

Fig. 13 Graphical representation of cloud service specifications implemented in angular

Similar to the model-based approach of Sect. 4.2, all the specifications, including their values, become instances for the related ontology classes as soon as the web page is submitted. Hence, the instance model is formally grounded with an ontology. Both the web-based and the model-based representations refer to the same ontology-based metamodel.

5 Conclusion

With ontology-based metamodels as described in this chapter it is possible to define the semantics of domain-specific modeling languages using ontologies. Models represented in such a modeling language are instances of ontology classes. Because of the formal representation, the models can be interpreted by software systems. On the other hand, the metamodels can be extended with graphical notations for the different modeling elements. Graphical modeling is more appropriate for humans than the creation of formal models. Thus, the ontology-based modeling approach allows for easy modeling using graphical notation and at the same time creates models that can be used for automation and machine-based interpretation. The ontology-based metamodeling was implemented by extending BPMN in a graphical modeling environment and as a web interface.

In future research, a modeling tool will be developed that integrates modeling and metamodeling in a graphical interface. A knowledge engineer can extend the domain-specific modeling language with new modeling elements on the spot and use them immediately.

Acknowledgements This research has received funding from the European Community's Framework Programme for Research and Innovation HORIZON 2020 (ICT-07-2014) under grant agreement number 644690 (CloudSocket).

References

APQC (2014) Process classification framework version 6.1.1

Azzini A, Braghin C, Damiani E, Zavatarelli, F (2013) Using semantic lifting for improving process mining: a data loss prevention system case study. SIMPDA, pp 62–73

C-SIG (2014) Cloud service level agreement standardization guidelines. EC Cloud Select Industry Group

De Angelis G, Pierantonio A, Polini A, Re B, Thönssen B, Woitsch R (2016) Modeling for learning in public administrations—the learn PAd approach. In: domain-specific conceptual modeling, Cham: Springer International Publishing, pp 575–594. https://doi.org/10.1007/978-3-319-39417-6_26

Dietz JLG (2006) Enterprise ontology. theory and methodology. Springer, Berlin Heidelberg

Emmenegger S, Hinkelmann K, Laurenzi E, Thönssen B, Witschel HF, Zhang C (2016). Workplace learning—providing recommendations of experts and learning resources in a context-sensitive and personalized manner. In: MODELSWARD 2016, Special session on learning modeling in complex organizations. Rome

Fill H-G, Karagiannis D (2016) On the conceptualisation of modelling methods using the ADOxx meta modelling platform. Enterp Model Inf Syst Architect - Int J Conceptual Mode 8(1): 4–25

Fill H-G, Schremser D, Karagiannis D (2013) A generic approach for the semantic annotation of conceptual models using a service-oriented architecture. Int J Knowled Manag 9(1):76–88. https://doi.org/10.4018/jkm.2013010105

Fowler M (2011) Domain-specific languages. Addison-Wesley, Upper Saddle River

Frank U (2010) Outline of a method for designing domain-specific modelling languages. University of Duisburg Essen, ICB

Gray J, Fisher K, Consel C, Karsai G, Mernik M, Tolvanen JP (2008) DSLs: the good, the bad, and the ugly. In: Conference on object oriented programming systems languages and applications archive. Nashville and {É}tats-Unis: ACM

Hinkelmann K, Gerber A, Karagiannis D, Thoenssen B, van der Merwe A, Woitsch R (2016a) A new paradigm for the continuous alignment of business and IT: Combining enterprise architecture modelling and enterprise ontology. Comput Ind 79: 77–86. https://doi.org/10.1016/j.compind. 2015.07.009

Hinkelmann K, Kritikos K, Kurjakovic S, Lammel B, Woitsch R (2016b) A modelling environment for business process as a service. CAiSE 2016: Advanced Information Systems Engineering Workshops, Ljubljana, Slovenia, pp 181–192

Hinkelmann K, Kurjakovic S, Lammel B, Laurenzi E, Woitsch R (2016c) A semantically-enhanced modelling environment for business process as a service. In: Fourth international conference on enterprise systems ES2016, Melbourne, Australia, 2–3 November 2016

Höfferer P (2007) Achieving business process model interoperability using metamodels and ontologies. In: European conference on information systems. university of St. Gallen. (pp 1620–1631). http://www.dke.at/fileadmin/DKEHP/publikationen/metamodell/Hoefferer_BP_interoperability_ontologies.pdf

Hrgovcic V, Karagiannis D, Woitsch R (2013). Conceptual modeling of the organisational aspects for distributed applications: the semantic lifting approach. In: COMPSACW 2013, 2013 IEEE 37th annual computer software and applications conference workshops, pp 145–150. IEEE. https://doi.org/10.1109/compsacw.2013.17

Hudak P, Paul (1996) Building domain-specific embedded languages. ACM Comput Surveys, 28(4es), 196–es. https://doi.org/10.1145/242224.242477

Kappel G, Kapsammer E, Kargl H, Kramler G, Reiter T, Retschitzegger W et al (2006) Lifting meta-models to ontologies: a step to the semantic integration of modeling languages. In: Nierstrasz O, Whittle J, Harel D, Reggio G (Eds.), Model driven engineering languages and systems, Proceedings of the 9th international conference, MoDELS 2006 (LNCS 4199, pp 528–542). Genova, Italy: Springer

Karagiannis D, Kühn H (2002) Metamodelling platforms. In: Bauknecht K, Min Tjoa A, Quirch-mayer G (Eds.), Proceedings of the third international conference EC-Web at DEXA 2002. Berlin: Springer

Karagiannis D, Woitsch R (2010) Knowledge Engineering in Business Process Management. Hand-book on business process management 2. Springer, Berlin Heidelberg, pp 463–485

Kelly S, Tolvanen J-P (2008) Domain-specific modeling: Enabling full code generation. Wiley, Hoboken

Kramler G, Kappel G, Reiter T, Kapsammer E, Retschitzegger W, Schwinger W (2006) Towards a semantic infrastructure supporting model-based tool integration. In GaMMa'06: Proceedings of the 2006 international workshop on global integrated model management (pp 43–46). New York, NY, USA: ACM Press

Jonkers H, Van Buuren R, Arbab F, De Boer F, Bonsangue M, Iacob M, Enschede AN (2003). Towards a language for coherent enterprise architecture descriptions. https://pdfs. semanticscholar.org/546f/0891738f53a6639e863454d915a71094d9ce.pdf

Laurenzi E, Hinkelmann K, Reimer U, Van Der Merwe A, Sibold P, Endl R (2017). DSML4PTM: a domain-specific modelling language for patient transferal management. In ICEIS 2017—Pro-ceedings of the 19th international conference on enterprise information systems vol. 3

Liao Y, Lezoche M, Panetto H, Boudjlida N, Loures ER (2015) Semantic annotation for knowledge explicitation in a product lifecycle management context: a survey. Comput Ind 71:24–34

Mernik M, Heering J, Sloane AM (2005) When and how to develop domain-specific languages. ACM Comput Surv 37(4):316–344. https://doi.org/10.1145/1118890.1118892

Nikles S, Brander S (2009) Separating conceptual and visual aspects in meta-modeling. In: Gerber A, Hinkelmann K, Kotze P, Reimer U, van der Merwe A (Eds.), Workshop on advanced enterprise architecture and repositories, Milano

OMG (2011) Business Process Model and Notation (BPMN) version 2.0. Needham, MA: object management group OMG. http://www.omg.org/spec/BPMN/2.0/PDF/

OMG (2014). OMG Meta Object Facility (MOF) Core Specification Version 2.4.2 (Vol. 2)

OMG (2016) Decision Model and Notation (DMN) V1.1. Object management group OMG. http://www.omg.org/spec/DMN/1.1

van Deursen A, Klint P, Visser J (2000) Domain-specific languages: an annotated bibliography. SIGPLAN Not 35(6):26–36. https://doi.org/10.1145/352029.352035

Woitsch R, Hinkelmann K, Juan Ferrer AM, Yuste JI (2016) Business Process as a Service (BPaaS): the smart BPaaS design environment. CAiSE 2016 industry track CEUR workshop proceedings, vol 1600. http://ceur-ws.org/Vol-1600, Ljubljana, Slovenia

W3C (2014). RDF Schema 1.1

Part IV
Artificial Intelligence

Searching and Browsing in Historical Documents—State of the Art and Novel Approaches for Template-Based Keyword Spotting

Michael Stauffer, Andreas Fischer and Kaspar Riesen

Abstract In many public and private institutions, the digitalization of handwritten documents has progressed greatly in recent decades. As a consequence, the number of handwritten documents that are available digitally is constantly increasing. However, accessibility to these documents in terms of browsing and searching is still an issue as automatic full transcriptions are often not feasible. To bridge this gap, *Keyword Spotting (KWS)* has been proposed as a flexible and error-tolerant alternative to full transcriptions. KWS provides unconstrained retrievals of keywords in handwritten documents that are acquired either *online* or *offline*. In general, offline KWS is regarded as the more difficult task when compared to online KWS where temporal information on the writing process is also available. The focus of this chapter is on handwritten historical documents and thus on offline KWS. In particular, we review and compare different state-of-the-art as well as novel approaches for *template-based* KWS. In contrast to *learning-based* KWS, template-based KWS can be applied to documents without any a priori learning of a model and is thus regarded as the more flexible approach.

Keywords Handwritten keyword spotting · Graph representation · Bipartite graph matching · Ensemble methods

M. Stauffer (✉) · K. Riesen
Institute for Information Systems, University of Applied Sciences and Arts
Northwestern Switzerland, Riggenbachstrasse 16, 4600 Olten, Switzerland
e-mail: michael.stauffer@fhnw.ch

K. Riesen
e-mail: kaspar.riesen@fhnw.ch

M. Stauffer
Department of Informatics, University of Pretoria, Pretoria 0083, South Africa

A. Fischer
Department of Informatics, University of Fribourg, 1700 Fribourg, Switzerland
e-mail: andreas.fischer@unifr.ch

A. Fischer
Institute of Complex Systems, University of Applied Sciences and Arts
Western Switzerland, 1705 Fribourg, Switzerland

© Springer International Publishing AG 2018
R. Dornberger (ed.), *Business Information Systems and Technology 4.0*,
Studies in Systems, Decision and Control 141,
https://doi.org/10.1007/978-3-319-74322-6_13

1 Broad Perspective and Outline

In the last decades, handwritten documents have become increasingly available digitally in many fields and applications. However, automatic full transcriptions of handwritten documents are far from perfect, especially as recognition is often negatively affected by degraded documents and/or different writing styles (Wicht et al. 2016). Thus, accessibility to handwritten documents with respect to browsing and searching is still an open issue. In order to overcome the obstacles of a full transcription, *Keyword Spotting (KWS)* has been proposed as a more error-tolerant and flexible approach for speech (Rose and Paul 1990), printed (Agazzi 1994), and handwritten documents (Manmatha et al. 1996). KWS refers to the task of retrieving any instance of a given query word in a particular document. In the case of historical handwritten documents, KWS is inherently an *offline* task, and as such, more complex than *online* KWS where temporal information on the writing process is also available. Since the focus of this chapter is on historical documents, only offline KWS—referred to as KWS from now on—can be applied.

1.1 Template-Based Versus Learning-Based KWS

Most KWS approaches are either *template-based* or *learning-based* algorithms. The following paragraphs provide a brief survey of methods stemming from both categories.

The earliest template-based KWS approaches are based on pixel-by-pixel matchings of word images (Manmatha et al. 1996). That is, the pixels of the word images are matched on the basis of Euclidean distance measures or affine transformations by the *Scott and Longuet*-algorithm (Scott and Longuet-Higgins 1991). Likewise, *Zones of Interest*, rather than single pixels, are matched in Leydier et al. (2007). More recently, word images have been described by binary features, so called *Gradient, Structural and Convexity (GSC)* features, and matched by correlation-like measures (Zhang et al. 2003).

However, single features tend to be affected by noise, and thus, more recent approaches to template-based KWS are based on matching sequences of feature vectors. These sequences of feature vectors are often used to represent certain characteristics of word images, such as, for example, projection profiles (Manmatha and Rath 2003; Rath and Manmatha 2003; Zhang et al. 2003), contours (Adamek et al. 2006; Can and Duygulu 2011), or geometrical characteristics (Marti and Bunke 2001; Manmatha and Rath 2003). However, more generic image feature descriptors have also been applied, for example, *Gabor* (Cao and Govindaraju 2007), *Histograms of Oriented Gradients* (Rodríguez-Serrano and Perronnin 2008; Terasawa and Tanaka 2009; Kovalchuk et al. 2014), *Local Binary Patterns* (Kovalchuk et al. 2014; Dey et al. 2016) or *Scale-Invariant Feature Transform* (Konidaris et al. 2015), to mention just a few. In a recent paper (Wicht et al. 2016), features are extracted by means of a *Convolutional Deep Belief Network*. Regardless of the employed feature descriptor, *Dynamic Time Warping (DTW)* is probably the most widely used method for matching sequences of features vectors and is actually used in various KWS publications (Marti and Bunke 2001; Manmatha and Rath 2003; Adamek et al. 2006; Frinken et al. 2012; Wicht et al. 2016).

In contrast to template-based approaches, learning-based KWS is based on statistical models that have to be trained a priori with respect to the actual spotting task on a (relatively large) training set of word or character images. Early approaches to learning-based KWS are based on *generalized Hidden Markov Models (gHMM)* that are trained on character images, i.e. images of Latin (Edwards et al. 2004) or Arabic (Chan et al. 2006) characters. However, character-based segmentations are often error-prone. Thus, more recent approaches are based on feature vectors of word images (Lavrenko et al. 2004), which are processed, for example, by means of *Continuous-HMM* (Rodríguez-Serrano and Perronnin 2009) or *Semi-Continuous-HMM* (Rodríguez-Serrano and Perronnin 2009, 2012), i.e. HMMs with a shared set of *Gaussian Mixture Models*. In Perronnin and Rodríguez-Serrano (2009), a *Fisher Kernel* is employed in conjunction with HMMs, while a line-based and lexicon-free HMM-approach is introduced in Fischer et al. (2012). In recent papers, HMMs were applied in combination with *Bag-of-Features* (Rothacker et al. 2013; Rothacker and Fink 2015), or *Deep Neural Networks* (Thomas et al. 2014; Wicht et al. 2016). Other learning-based KWS approaches are for example based on *Support Vector Machines* (Huang et al. 2011; Almazán et al. 2014), or *Neural Networks* (Aghbari and Brook 2009; Frinken et al. 2012), to name just two examples.

Generally, learning-based approaches result in higher KWS accuracy when compared to template-based approaches. However, this advantage is accompanied by a loss of flexibility, which is due to the need to learn the parameters of the actual model. In particular, template-based KWS is independent of both the actual representation formalism and the language of the underlying document.

1.2 Statistical Versus Structural Representation

All of the KWS methodologies mentioned so far are based on statistical representation formalisms (this accounts for both template-based and learning-based methods). That is, word images or subimages are represented by means of feature vectors or sequences of feature vectors encoding certain local or global characteristics. However, in recent years a tendency towards structural representation formalisms has been observed in various fields of pattern recognition (Conte et al. 2004; Foggia et al. 2014; Riesen 2015; Stauffer et al. 2017d). Structural representations such as strings, trees, or graphs (whereby strings and trees can be seen as special cases of graphs) are more sophisticated data structures when compared to vectorial formalisms. In contrast to feature vectors, graphs are able to adapt both their size and structure to the underlying pattern. Moreover, graphs are able to represent binary relationships that might exist between the subparts of the represented pattern. This turns graphs into a natural and comprehensive way for representing handwriting.

Given the power and flexibility of graphs, it might be rather surprising that few graph-based KWS approaches have been proposed until now (Wang et al. 2014; Riba et al. 2015; Bui et al. 2015; Stauffer et al. 2016b). One possible reason for this observation is the general increase in the complexity of many algorithms that use graphs rather than vectors as their input.

The first graph-based KWS approach was introduced in Wang et al. (2014), where certain keypoints in word images are represented by nodes, while edges are used to represent strokes between selected keypoints. The matching procedure is then conducted in two stages. First, graph dissimilarities between pairs of subgraphs are computed by means of a fast approximation algorithm (Riesen and Bunke 2009). Secondly, an optimal cost assignment is found by means of DTW. In Bui et al. (2015) and Riba et al. (2015), two similar approaches are shown, where nodes represent prototype strokes, while edges are used to represent the connectivity between strokes. Finally, graph dissimilarities are computed by the same algorithm as in Wang et al. (2014). One of the most recent graph-based KWS approaches was proposed by Stauffer et al. (2016a, b), where four different graph representation formalisms are introduced and compared with each other.

1.3 Outline

The present chapter focuses on reviewing template-based approaches for offline KWS. In particular, we review different state-of-the-art and novel approaches for template-based KWS for both statistical and structural representation formalisms in Sects. 2 and 3, respectively. Section 4 deals with an empirical comparison of both representations of two historical benchmark documents. Finally, Sect. 5 concludes this chapter and outlines future trends and rewarding opportunities.

2 Statistical Template-Based Keyword Spotting

In this section we review four different DTW-based systems for template-based KWS based on statistical representations, viz. Marti and Bunke (2001) (termed DTW'01), Rodríguez-Serrano and Perronnin (2008) (termed DTW'08), Terasawa and Tanaka (2009) (termed DTW'09), and Wicht et al. (2016) (termed DTW'16).

Basically, the four reviewed KWS systems consist of three subsequent steps, as illustrated in Fig. 1. First, document images are preprocessed (A) in order to minimise variations caused, for instance, by noisy background images, skewed scanning, or degraded documents. Subsequently, document images are automatically segmented into word images. Based on preprocessed word images, sequences of feature vectors are extracted by means of different feature descriptors (B). Finally, a query word

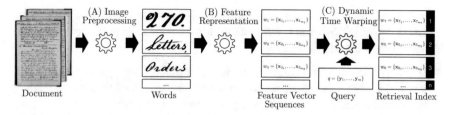

Fig. 1 Process of statistical template-based keyword spotting

(represented by a sequence of feature vectors) undergoes pairwise matching with a set of document words (represented by a set of sequences of feature vectors) (C). A retrieval index for a queried keyword can be derived on the basis of these dissimilarities. In the best possible case, this index represents all n instances of a given query word as the top-n results.

These three steps are described in greater detail in the following three subsections. It should be noted that the four systems only differ with respect to the extracted features. That is, the image preprocessing as well as the DTW-matching is conducted in quite a similar way in all four approaches.

2.1 Image Preprocessing

Image preprocessing aims at reducing variations caused by different writing styles (i.e. interpersonal variations) as well as the document itself (e.g. pixel noise, skewed scanning, or degraded documents). The reviewed systems rely on the following preprocessing steps.

The first preprocessing step addresses the issue of noisy background (e.g. by enhancing edges by a *Difference of Gaussians* (Fischer et al. 2010)). Next, document images are binarized by a global threshold and automatically segmented into single word images or word fragments. In addition, the skew, i.e. the inclination of the document, is also estimated on the lower baseline of a line of text and then corrected on the documents or single word images (Marti and Bunke 2001). Finally, the slant, i.e. the inclination of the handwriting, is also removed using a shear transformation (Marti and Bunke 2001).

2.2 Feature Representation

Based on preprocessed and segmented word images, sequences of feature vectors $\{x_1, \ldots, x_m\}$ are extracted by means of a sliding window approach. In particular, a sliding window (with a user-defined width) is seamlessly moved over a word image from left to right, and thus, one feature vector $\mathbf{x}_i$ is extracted at each window position i. The different DTW-based KWS systems differ with respect to the actual features extracted from the sliding window.

- **DTW'01** (Geometrical Features): In Marti and Bunke (2001), nine different geometrical features are defined for each window position. The first group of features describes the sliding window from a global perspective by the weight, center, and second order moment of the sliding window. Four features describe the position and orientation of the upper and lower contour in the sliding window, respectively. The two remaining features are used to characterize the number of black-white transitions in the vertical direction, as well as the number of black pixels between the upper and lower contour.
- **DTW'08 and DTW'09** (Histogram of Oriented Gradient (HoG) Features): In Rodríguez-Serrano and Perronnin (2008), HoG-features are locally extracted at each window position. In particular, the window is split into $M \times N$ cells of equal size. Based on the horizontal and vertical gradient components, the gradient magnitude m and angle θ are computed for each foreground pixel in the window cell. Thus, the gradient angles can serve to create a histogram with T radial bins. Angle θ determines the closest bin, while m sums up the corresponding bin. Hence, $M \times N \times T$ features are extracted for each window position. In Terasawa and Tanaka (2009), similar HoG-like features are extracted for overlapping blocks of cells rather than single cells.
- **DTW'16** (Deep Learning Features): In Wicht et al. (2016), a *Convolutional Deep Belief Network* based on two *Convolutional Restricted Boltzmann Machines (CRBM)* is used to extract features at each window position. In particular, the network is trained in an unsupervised manner in two subsequent steps. First, an image of the sliding window is used to train the first CRPM. The output of this layer is reduced by a pooling layer and used as input for the training of the second CRBM. Finally, the output of the second CRPM is again reduced by a pooling layer and used as a feature vector.

2.3 Dynamic Time Warping

All of the keyword spotting systems reviewed are based on matching a query word q with all document words $w_i \in \{w_1, \ldots, w_N\}$ by means of the dynamic programming approach DTW. In particular, DTW optimally aligns two sequences of features vectors $X = \{\mathbf{x}_1, \ldots, \mathbf{x}_m\}$ and $Y = \{\mathbf{y}_1, \ldots, \mathbf{y}_n\}$ representing a query word q and a specific document word w_i along one common time axis using a dynamic programming approach. The alignment cost between each pair of feature vectors $(\mathbf{x}, \mathbf{y}) \in \mathbb{R}^k \times \mathbb{R}^k$ is given by the squared Euclidean distance. Formally,

$$d(\mathbf{x}, \mathbf{y}) = \sum_{i=1}^{k} (\hat{x}_i - \hat{y}_i)^2, \tag{1}$$

where k denotes the number of features, and $\hat{x}_i$ and $\hat{y}_i$ denote features normalized with a z-score. The DTW distance $D(q, w)$ between two sequences of feature vectors is then given by the minimum alignment cost found by dynamic programming. Formally,

$$D(X, Y) = \sum_{k=1}^{K} d(\mathbf{x}_{i_k}, \mathbf{y}_{j_k}),$$ (2)

where K is the length of the optimal warping path $((i_1, j_1), \ldots, (i_K, j_K))$ (Rath and Manmatha 2003). A *Sakoe-Chiba band* that constrains the warping path is often applied to speed up this procedure (Sakoe and Chiba 1978). Finally, a retrieval index can be created based on DTW distances between a query and all document words.

3 Structural Template-Based Keyword Spotting

In this section, we review two graph-based systems proposed by the authors of the present chapter for template-based KWS based on structural representations (Stauffer et al. 2016b, 2017a). Similarly to the statistical systems described in Sect. 2, the graph-based approaches consist of three subsequent steps, as illustrated in Fig. 2. First, document images are preprocessed and segmented into single word images (A). On the basis of preprocessed word images, graphs are extracted by means of a graph extraction algorithm (B). The actual keyword spotting is then based on a pairwise matching of a query graph with the set of all document graphs (C). A retrieval index is finally derived based on the resulting graph dissimilarities. In the following three subsections these steps are described in greater detail.

3.1 Image Preprocessing

The image preprocessing is based on similar steps as described in Sect. 2.1. That is, document images are filtered and binarized (Fischer et al. 2010), automatically segmented into word images, and manually corrected, if necessary. Next, the skew

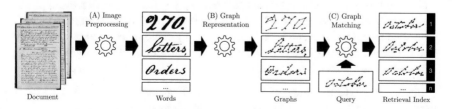

Fig. 2 Process of structural template-based keyword spotting

is estimated on lines of text and corrected on single words (Marti and Bunke 2001). However, in contradiction to the process described in Sect. 2.1, the slant is not corrected. Finally, word images are skeletonized by a 3×3 thinning operator (Guo and Hall 1989).

3.2 Graph Representation

In Stauffer et al. (2016b, 2017a), graphs serve to represent preprocessed and segmented word images. A graph g is defined as a four-tuple $g = (V, E, \mu, \nu)$ where V and E are finite sets of nodes and edges, and $\mu : V \rightarrow L_V$ as well as $\nu : E \rightarrow L_E$ are labeling functions for nodes and edges, respectively. All of the following four graph extraction algorithms (originally presented in Stauffer et al. (2016a)) result in graphs where nodes are labeled with two-dimensional numerical labels, while edges remain unlabeled, i.e. $L_V = \mathbb{R}^2$ and $L_E = \{\}$.

- **Keypoint**: The first graph extraction algorithm makes use of keypoints in word images such as start, end, and junction points. These keypoints are represented as nodes that are labeled with the corresponding (x, y)-coordinates. Between pairs of keypoints further intermediate points are converted to nodes and added to the graph at equidistant intervals. Finally, undirected edges are inserted into the graph for each pair of nodes that is directly connected by a stroke.
- **Grid**: The second graph extraction algorithm is based on a grid-wise segmentation of word images. For each segment, a node is inserted into the graph and labeled by the (x, y)-coordinates of the center of mass of this segment. Undirected edges are inserted between two neighboring segments that are actually represented by a node. Lastly, the inserted edges are reduced by means of a *Minimal Spanning Tree* algorithm (Kruskal 1956).
- **Projection**: The next graph extraction algorithm works in a similar way as Grid. However, this method is based on an adaptive segmentation of word images by means of horizontal and vertical projection profiles. A node is inserted into the graph for each segment and labeled by the (x, y)-coordinates of the corresponding center of mass. Undirected edges are inserted into the graph for each pair of nodes that is directly connected by a stroke in the original word image.
- **Split**: The last graph extraction algorithm is based on an iterative segmentation of word images. That is, segments are iteratively split into smaller subsegments until the width and height of all segments is below a certain threshold. A node is inserted into the graph and labeled by the (x, y)-coordinates of the point closest to the center of mass of each segment. For the insertion of the edges, a similar procedure as for Projection is applied.

Finally, the dynamic range of the (x, y)-coordinates of each node label $\mu(v)$ is normalized with a z-score (regardless the extraction algorithm). Formally,

$$\hat{x} = \frac{x - \mu_x}{\sigma_x} \text{ and } \hat{y} = \frac{y - \mu_y}{\sigma_y}, \tag{3}$$

where (μ_x, μ_y) and (σ_x, σ_y) represent the mean and standard deviation of all (x, y)-coordinates in the graph under consideration.

3.3 Graph Matching

The actual keyword spotting is based on pairwise matching of a query graph q with all document graphs $w_i \in \{w_1, \ldots, w_n\}$. Several approaches for graph matching have been proposed (Conte et al. 2004; Foggia et al. 2014). However, *Graph Edit Distance (GED)* is widely accepted as one of the most flexible and powerful paradigms available (Bunke and Allermann 1983). Given a query graph q and a document graph w, the basic idea of GED is to transform q into w using a sequence of edit operations. A standard set of edit operations is given by *insertions*, *deletions*, and *substitutions* of both nodes and edges. A set $\{e_1, \ldots, e_k\}$ of k edit operations e_i that transform q completely into w is referred to as an *edit path* $\lambda(q, w)$ between q and w.

To find the most suitable edit path, a domain-specific cost function $c(e)$ is usually introduced for each edit operation e. This cost function is used to measure the degree of deformation of a given edit operation. Given an adequate cost model, the graph edit distance $d_{\mathrm{GED}}(q, w)$, or d_{GED} for short, between q and w is defined by

$$d_{\mathrm{GED}}(q, w) = \min_{\lambda \in \Upsilon(q,w)} \sum_{e_i \in \lambda} c(e_i), \tag{4}$$

where $\Upsilon(q, w)$ denotes the set of all edit paths between q and w.

For the exact computation of d_{GED}, it is common to employ A*-based search techniques using heuristics (Fankhauser et al. 2011). However, these exhaustive search procedures are exponential with respect to the number of nodes of the involved graphs. Hence, in Stauffer et al. (2017a) the *Bipartite Graph Matching (BP)* algorithm (Riesen and Bunke 2009) is used, which approximates the GED in cubic time. Based on the resulting suboptimal graph edit distance $d_{\mathrm{BP}}(q, w)$, a retrieval index is computed between query and document words.

3.3.1 Ensemble Methods

In Stauffer et al. (2017a), all graph representations as introduced in Sect. 3.2 are used in one KWS system at the same time. This approach is a well-known strategy from the field of *multiple classifier systems*, also referred to as *ensemble methods*. In particular, several query graphs (representing the same query word) are matched with several document graphs (representing the same document word). Next, different strategies are applied to combine the individual graph edit distances (derived

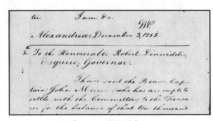
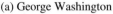
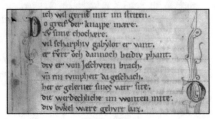

(a) George Washington   (b) Parzival

Fig. 3 Exemplary excerpts of the two datasets

from the different representations). In Stauffer et al. (2017a), the minimal (termed min), maximal (termed max), or mean (termed mean) graph edit distance is used to condense the multiple distances to one retrieval index. Moreover, the most promising individual graph representations presented in Stauffer et al. (2016a), viz. `Keypoint` and `Projection`, are used to derive two weighted sums (termed sum_α and sum_{map}). The former sum makes use of a user-defined weighting value while the latter is based on a relative weighting that relies on the *Mean Average Precision* of the individual ensemble members. Eventually, the two normalized distances are summed up to form one single retrieval index.

4 Experimental Evaluation

In this section, we compare the reviewed statistical and structural approaches for template-based KWS with each other. The optimal parameters of the systems are taken from the corresponding papers. The comparison is carried out on two historical documents, viz. the *George Washington letters (GW)* and the *Parzival manuscript (PAR)* as shown in Fig. 3. GW is based on letters that are written in English and consists of twenty pages with a total of 4,894 handwritten words.[1] Variations caused by both degradation and writing style are low. PAR is based on a manuscript that is written in Middle High German and consists of 45 pages with a total of 23,478 handwritten words.[2] There are marked variations caused by degradation, while variations caused by writing style are low.

The performance of all KWS systems is measured by the *Recall (R)* and *Precision (P)*

$$R = \frac{TP}{TP + FN} \text{ and } P = \frac{TP}{TP + FP}, \tag{5}$$

---

[1] George Washington Papers at the Library of Congress, 1741–1799: Series 2, Letterbook 1, pp. 270–279 & 300–309, http://memory.loc.gov/ammem/gwhtml/gwseries2.html.

[2] Parzival at IAM historical document database, http://www.fki.inf.unibe.ch/databases/iam-historical-document-database/parzival-database.

which are both based on the number of *True Positives (TP)*, *False Positives (FP)*, and *False Negatives (FN)*.

Both recall and precision can be computed for two types of thresholds, viz. *local* and *global* thresholds. In the case of global thresholds, the quality of the KWS system is measured by *Average Precision (AP)*, which is the area under the *Recall-Precision (RP)* curve for all keywords given a single (global) threshold. In the case of local thresholds, the performance is indicated by *Mean Average Precision (MAP)*, that is the mean over the AP of each individual keyword query. Generally, global thresholds are regarded as the more realistic and challenging scenario.

For both benchmark datasets, the MAP and AP are given in Table 1. First, we compare the three individual approaches independently of each other (statistical, structural, and structural ensemble). In the case of statistical KWS, we observe that DTW'16 is the best approach in three out of four cases. However, DTW'09 also outperforms the two other statistical approaches, especially on PAR. In the case of structural KWS, we observe that Keypoint results in the highest KWS accuracy on GW, while Projection achieves the highest accuracy on PAR. On both datasets Grid and Split result in the lowest accuracy when compared to all other graph extraction methods. In the case of graph-based ensemble methods, we observe that the ensemble strategy mean achieves the best result in two out of four cases and the second and third best result in two cases.

Table 1 Mean average precision (MAP) using local thresholds and average precision (AP) using a global threshold for all DTW- and graph-based KWS systems. The first, second, and third best systems are indicated by (1), (2), and (3)

Method	GW		PAR	
	MAP	AP	MAP	AP
DTW				
DTW'01	45.26	33.24	46.78	50.67
DTW'08	63.39	41.20	47.52	55.82
DTW'09	64.80	43.76	73.49	69.10
DTW'16	68.64	56.98 (3)	72.38	72.71 (3)
Graph (Single)				
Keypoint	66.08	55.22	62.04	60.76
Grid	60.02	46.09	56.50	46.00
Projection	61.43	49.34	66.23	62.38
Split	60.23	48.08	59.44	56.25
Graph (Ensemble)				
min	70.56 (1)	56.82	67.90	62.33
max	62.58	47.94	67.57	50.59
mean	69.16 (3)	57.11 (2)	79.38 (1)	73.77 (1)
sum_α	68.44	55.78	74.51 (3)	68.12
sum_{map}	70.20 (2)	57.38 (1)	76.80 (2)	73.56 (2)

Comparing all systems with each other, we observe that the graph-based ensemble methods achieve the overall best results on both datasets and with both thresholds (with statistical significance (t-test, $\alpha = 0.05$)). In particular, the ensemble mean and sum_{map} outperform all other statistical and structural KWS approaches. This is particularly interesting as the DTW-based systems (Terasawa and Tanaka 2009; Wicht et al. 2016) use advanced feature sets, while the graph-based methods rely on coordinate labels only.

5   Conclusion and Outlook

In this chapter, different approaches for template-based Keyword Spotting (KWS) are reviewed. These methods basically differ in the formalism used to represent handwriting, viz. by means of statistical or structural representations. That is, preprocessed and segmented word images are either represented as sequences of feature vectors (in the case of statistical KWS) or graphs (in the case of structural KWS). The actual keyword spotting is then based on a matching of a query word with all document words by a dynamic programming approach or graph matching, respectively.

For the experimental evaluation both statistical and structural KWS approaches are compared with each other on two different benchmark datasets, viz. George Washington (GW) and Parzival (PAR). In the case of statistical KWS, DTW'16 is to favour on both datasets in most of the cases. In the case of structural methods, we observe that either `Keypoint` or `Projection` result in the highest accuracy on GW and PAR, respectively. Moreover, we observe that graph-based ensemble methods are able to clearly outperform all individual methods, as well as all statistical approaches.

One might argue that graph-based approaches are limited by the increased complexity of the matching procedure when compared to statistical approaches. However, recent papers (e.g. Stauffer et al. 2017b, c; Ameri et al. 2017) show that the complete KWS procedure with graphs can be substantially speeded up by filters and other heuristics. This makes graphs a versatile alternative for template-based KWS.

In future work, we see great potential in the combination of statistical and structural approaches. For instance, we plan to combine the matching scores derived by matching subgraphs of a sliding window with a DTW-based approach.

Acknowledgements This work has been supported by the Hasler Foundation Switzerland.

References

Adamek T, O'Connor NE, Smeaton AF (2006) Word matching using single closed contours for indexing handwritten historical documents. Int J Doc Anal Recogn 9(2–4):153–165

Agazzi O (1994) Keyword spotting in poorly printed documents using pseudo 2-D hidden Markov models. IEEE Trans Pattern Anal Mach Intell 16(8):842–848

Aghbari ZA, Brook S (2009) HAH manuscripts: a holistic paradigm for classifying and retrieving historical Arabic handwritten documents. Expert Syst Appl 36(8):10942–10951

Almazán J, Gordo A, Fornés A, Valveny E (2014) Segmentation-free word spotting with exemplar SVMs. Pattern Recogn 47(12):3967–3978

Ameri M, Stauffer M, Riesen K, Bui T, Fischer A (2017) Keyword spotting in historical documents based on handwriting graphs and Hausdorff edit distance. In: International graphonomics society conference

Bui QA, Visani M, Mullot R (2015) Unsupervised word spotting using a graph representation based on invariants. In: International conference on document analysis and recognition, pp 616–620

Bunke H, Allermann G (1983) Inexact graph matching for structural pattern recognition. Pattern Recogn Lett 1(4):245–253

Can EF, Duygulu P (2011) A line-based representation for matching words in historical manuscripts

Cao H, Govindaraju V (2007) Template-free word spotting in low-quality manuscripts. In: International conference on advances in pattern recognition, pp 1–5

Chan J, Ziftci C, Forsyth D (2006) Searching off-line arabic documents. IEEE Comput Soc Conf Comput Vis Pattern Recogn 2:1455–1462

Conte D, Foggia P, Sansone C, Vento M (2004) Thirty years of graph matching in pattern recognition. Int J Pattern Recogn Artif Intell 18(03):265–298

Dey S, Nicolaou A, Llados J, Pal U (2016) Local binary pattern for word spotting in handwritten historical document. Computing Research Repository

Edwards J, Teh YW, Bock R, Maire M, Vesom G, Forsyth DA (2004) Making latin manuscripts searchable using gHMM's. Int Conf Neural Inf Process Syst 17:385–392

Fankhauser S, Riesen K, Bunke H (2011) Speeding up graph edit distance computation through fast bipartite matching. In: Graph-based representations in pattern recognition, pp 102–111

Fischer A, Indermühle E, Bunke H, Viehhauser G, Stolz M (2010) Ground truth creation for handwriting recognition in historical documents. In: International workshop on document analysis systems, New York, USA, pp 3–10

Fischer A, Keller A, Frinken V, Bunke H (2012) Lexicon-free handwritten word spotting using character HMMs. Pattern Recogn Lett 33(7):934–942

Foggia P, Percannella G, Vento M (2014) Graph matching and learning in pattern recognition in the last 10 years. Int J Pattern Recogn Artif Intell 28(01)

Frinken V, Fischer A, Manmatha R, Bunke H (2012) A novel word spotting method based on recurrent neural networks. IEEE Trans Pattern Anal Mach Intell 34(2):211–224

Guo Z, Hall RW (1989) Parallel thinning with two-subiteration algorithms. Commun ACM 32(3):359–373

Huang L, Yin F, Chen QH, Liu CL (2011) Keyword spotting in offline chinese handwritten documents using a statistical model. In: International conference on document analysis and recognition, pp 78–82

Konidaris T, Kesidis AL, Gatos B (2015) A segmentation-free word spotting method for historical printed documents. Pattern Anal Appl

Kovalchuk A, Wolf L, Dershowitz N (2014) A simple and fast word spotting method. In: International conference on frontiers in handwriting recognition, pp 3–8

Kruskal JB (1956) On the shortest spanning subtree of a graph and the traveling salesman problem. Proc Am Math Soc 7(1):48–48

Lavrenko V, Rath T, Manmatha R (2004) Holistic word recognition for handwritten historical documents. In: International workshop on document image analysis for libraries, pp 278–287

Leydier Y, Lebourgeois F, Emptoz H (2007) Text search for medieval manuscript images. Pattern Recogn 40(12):3552–3567

Manmatha R, Han C, Riseman E (1996) Word spotting: a new approach to indexing handwriting. In: Computer vision and pattern recognition, pp 631–637

Manmatha R, Rath TM (2003) Indexing of handwritten historical documents—recent progress. In: Symposium on document image understanding technology, pp 77–85

Marti UV, Bunke H (2001) Using a statistical language model to improve the performance of an HMM-based cursive handwriting recognition systems. Int J Pattern Recogn Artif Intell 15(01):65–90

Perronnin F, Rodríguez-Serrano JA (2009) Fisher kernels for handwritten word-spotting. In: International conference on document analysis and recognition, pp 106–110

Rath T, Manmatha R (2003) Word image matching using dynamic time warping. In: Computer vision and pattern recognition, vol 2, pp II–521–II–527

Riba P, Llados J, Fornes A (2015) Handwritten word spotting by inexact matching of grapheme graphs. In: International conference on document analysis and recognition, pp 781–785

Riesen K (2015) Structural pattern recognition with graph edit distance. In: Advances in computer vision and pattern recognition, Cham

Riesen K, Bunke H (2009) Approximate graph edit distance computation by means of bipartite graph matching. Image Vis Comput 27(7):950–959

Rodríguez-Serrano JA, Perronnin F (2008) Local gradient histogram features for word spotting in unconstrained handwritten documents. In: International conference on frontiers in handwriting recognition, pp 7–12

Rodríguez-Serrano JA, Perronnin F (2009) Handwritten word-spotting using hidden Markov models and universal vocabularies. Pattern Recogn 42(9):2106–2116

Rodríguez-Serrano JA, Perronnin F (2012) A model-based sequence similarity with application to handwritten word spotting. IEEE Trans Pattern Anal Mach Intell 34(11):2108–20

Rose R, Paul D (1990) A hidden Markov model based keyword recognition system. In: IEEE international conference on acoustics, speech, and signal processing, pp 129–132

Rothacker L, Fink GA (2015) Segmentation-free query-by-string word spotting with bag-of-features HMMs. In: International conference on document analysis and recognition, pp 661–665

Rothacker L, Rusinol M, Fink Ga (2013) Bag-of-features HMMs for segmentation-free word spotting in handwritten documents. In: International conference on document analysis and recognition, pp 1305–1309

Sakoe H, Chiba S (1978) Dynamic programming algorithm optimization for spoken word recognition. IEEE Trans Acoust Speech, Signal Process 26(1):43–49

Scott GL, Longuet-Higgins HC (1991) An algorithm for associating the features of two images. Proc Roy Soc B: Biol Sci 244(1309):21–26

Stauffer M, Fischer A, Riesen K (2016a) A novel graph database for handwritten word images. In: International workshop on structural, syntactic, and statistical pattern recognition

Stauffer M, Fischer A, Riesen K (2016b) Graph-based keyword spotting in historical handwritten documents. In: International workshop on structural, syntactic, and statistical pattern recognition

Stauffer M, Fischer A, Riesen K (2017a) Ensembles for graph-based keyword spotting in historical handwritten documents. In: International conference on document analysis and recognition

Stauffer M, Fischer A, Riesen K (2017b) Speeding-up graph-based keyword spotting by quadtree segmentations. In: International conference on computer analysis of images and patterns

Stauffer M, Fischer A, Riesen K (2017c) Speeding-up graph-based keyword spotting in historical handwritten documents. In: Graph-based representations in pattern recognition

Stauffer M, Tschachtli T, Fischer A, Riesen K (2017d) A survey on applications of bipartite graph edit distance. In: Graph-based representations in pattern recognition

Terasawa K, Tanaka Y (2009) Slit style HOG feature for document image word spotting. In: International conference on document analysis and recognition, pp 116–120

Thomas S, Chatelain C, Heutte L, Paquet T, Kessentini Y (2014) A deep HMM model for multiple keywords spotting in handwritten documents. Pattern Anal Appl 18(4):1003–1015

Wang P, Eglin V, Garcia C, Largeron C, Llados J, Fornes A (2014) A novel learning-free word spotting approach based on graph representation. In: International workshop on document analysis systems, pp 207–211

Wicht B, Fischer A, Hennebert J (2016) Deep learning features for handwritten keyword spotting. In: International conference on pattern recognition

Zhang B, Srihari SN, Huang C (2003) Word image retrieval using binary features. In: Document recognition and retrieval, p 45

How to Teach Blockchain in a Business School

Walter Dettling

Abstract There are different approaches to developing a syllabus for a "blockchain curriculum" in a business school. The following chapter identifies many questions which can guide lecturers through their own process of creating a blockchain syllabus. The list of topics and figures presented indicate how much time could be spent on these topics. This list does not present a definite syllabus, because the essence of blockchain is that it is a holistic approach which goes beyond naming topics. It raises certain questions and is intended to start a broader discussion which will eventually lead to a systematic approach to this multidimensional challenge. Structuring this journey leads to the following four main questions, which are considered relevant in this discussion:

- What are the key research topics in blockchain?
- What is the impact of blockchain and by what criteria do we define impact? Which building blocks of blockchain are relevant to understanding the blockchain mechanism?
- Which applications (beside Bitcoin) should be discussed and are expected to become important in the near and distant future?

This journey closes with some methodological remarks based on my experience of teaching blockchain classes. Finally, it is obvious that any curriculum will be condemned to continuous adaption because blockchain technology is changing in an ongoing way.

Keywords Blockchain · Bitcoin · Syllabus · Understanding blockchain

W. Dettling (✉)
Institute for Information Systems, University of Applied Sciences and Arts Northwestern
Switzerland, Peter Merian-Strasse 86, 4002 Basel, Switzerland
e-mail: walter.dettling@fhnw.ch

© Springer International Publishing AG 2018
R. Dornberger (ed.), *Business Information Systems and Technology 4.0*,
Studies in Systems, Decision and Control 141,
https://doi.org/10.1007/978-3-319-74322-6_14

1 Introduction

For most people today, Bitcoin is a synonym for a digital currency or blockchain and, to a certain extent, they are, right about this. There are already many different implementations of a blockchain in existence, but it is still worth examining the starting point of Bitcoin, which is the first public and most successful blockchain application so far (Nakamoto 2008).

I am a mathematician who teaches financial mathematics and business information systems. Besides teaching I also work for a software company which develops and implements standard business software for medium-sized companies. I take two different approaches when developing the content of a teaching topic. As a math teacher, I normally start with the building elements of a theory or method and end with the final "proof" as to why and how something works. As a business information systems teacher, I literally start from the opposite direction: From the impact or relevance of a topic, I deduce what elements lead to the understanding of a tool or method.

Although I know that I am still at a basic level of the blockchain learning curve, I have tried to structure my experiences and the ideas I have come across. This structure, however, does not describe my own journey in its chronological order or reflect the initial questions I had during my own learning experience.

In 2016, I started teaching blockchain in seminars for executives who had been looking for a comprehensive introduction to this topic. The seminar was called "Digital business models with blockchain" and covered a simple introduction and different use cases of blockchain. I then considered what to teach students of business information technology and management about blockchain, because it makes a huge difference whether you teach software engineers or business students. Software engineers can be introduced to fundamental topics such as cryptography, hashing, Merkel trees, etc. and then go on to existing implementations of blockchains and their programming environments. For business students, it is harder to find a starting point. What are the fundamental topics business students need to know about in order to understand blockchain and furthermore, what do they need to know to evaluate blockchain technology?

At the blockchain consensus 2017 conference in New York, five members of academia, representing MIT, Duke University and the University of Nicosia in Cyprus, discussed the issue of teaching blockchain at a university (Castor 2017). The panellists had a controversial discussion about the role of universities in this field:

- Must universities develop a young workforce with blockchain skills for start-ups or should they instead teach fundamental skills like cryptography or finance and leave the topics on the frontline of development to the start-ups?
- How can a professor seriously prepare a blockchain lesson, when the research delivers new evidence almost daily?
- How closely should universities partner with companies interested in blockchain know-how or more precisely, in blockchain talent? The fact that there are not

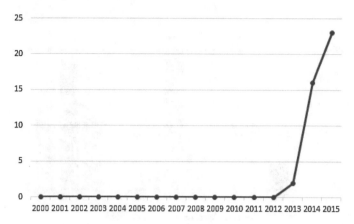

Fig. 1 Publication year of the selected primary papers. (*Source* Yli-Huumo et al. 2016, p. 9)

enough trained people on the market leads to strange effects: Students who have taken only one or two classes in this subject are hired as blockchain experts.

The panellists could not reach a consensus on these questions and so, not surprisingly, the conclusion of the discussion was: "Even academics can't keep pace with blockchain change".

2 What Are the Relevant Trends in Blockchain Research?

The inflow of new research articles about blockchain in my Mendeley account is constant. It would be presumptuous to assume that anybody can summarise the current research on blockchain. Yli-Huumo et al. (2016) published a review of the current research on blockchain technology. They carried out a systematic screening and analysis of scientific papers on technical research about blockchain, starting in the year 2000. After excluding non-relevant papers according to different criteria, they ended up with 41 primary papers extracted from scientific databases. Figure 1 shows the publication year distribution of the selected primary papers. All selected papers were published after the year 2012, which confirms again that blockchain is a very recent topic of research.

Another non-surprising finding of this study was that the majority of research papers (80.5%) were related to Bitcoin, and less than a fifth were not focused on Bitcoin.

A closer analysis on the types of papers in this study shows that the number of papers with an applied focus was increasing (see Fig. 2). In fact, Yli-Huumo et al. conclude that most of the papers published by the industry are not included in scientific databases, because industry papers can be found as white papers and are not often published in peer-reviewed conferences or journals.

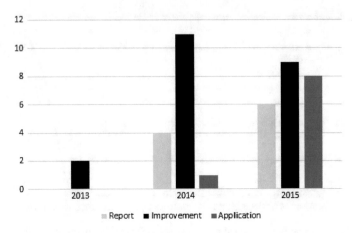

Fig. 2 Blockchain research paper types by year. (*Source* Yli-Huumo et al. 2016, p. 14)

This leads to the second challenge, that of defining relevant topics for students of business management.

3   What Is the Impact of Blockchain?

What are the trends and relevant topics related to blockchain outside academia? We can assume that students interested in this topic are more influenced by public discussions or the business-related impact than by academic research.

According to Tapscott and Tapscott (2016), blockchain technology is regarded as the most important driver for the development of business in the next decade: "Blockchain technology is complex, but the idea is simple….On the blockchain, trust is established, not by powerful intermediaries like banks, governments and technology companies, but through mass collaboration and clever code. Blockchains ensure integrity and trust between strangers….In other words, it's the first native digital medium for value, just as the Internet was the first native digital medium for information."

This very general approach stands for many similar publications in the business area. I abstain from citing here the white papers about blockchain by the big four auditing firms or similar organisations. They are doing a great job in supporting their clients in strategic thinking but, for several reasons, I avoid being influenced by their mind-set. I think our students should have the opportunity to study new topics in their native states. Or, to use a metaphor: Healthy studying is not only like eating basic and organic food but also learning to cook one's own tasty meals from recognized ingredients. With this experience, students are better prepared to decide and judge what they are offered from the huge field of processed brain food.

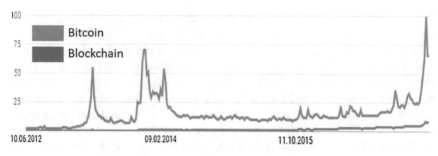

Fig. 3 Google Trends search rank of "Blockchain and Bitcoin", established June 10th, 2017

Using Google Trends as another indicator reveals a similar split in public interest as already stated in the case of academia (see Fig. 3). Bitcoin is the overwhelming popular topic and blockchain is lagging.

This corresponds to the interest of most students who are first and foremost motivated by the monetary aspect of bitcoin and not the underlying mechanism. It should be accepted that for now and some time Bitcoin is part of any blockchain curriculum. Not only because of its recent desirability but because Bitcoin is the first and still the most predominant implementation of a blockchain. The relevance of Bitcoin as an instance of blockchain is undermined by the fact that not only does the public use the terms Bitcoin and blockchain interchangeably, but also that there is no exact distinction between the two terms in academic research. Yli-Huumo et al. (2016) illustrate this matter in an exemplary manner. In their summary of blockchain challenges, most of the topics which have been discussed in the last two years relate to Bitcoin trading and mining. There is obviously no other use case which leads to so many questions and discussions about unsolved blockchain problems. These problems are new for everybody because this kind of blockchain did not exist until 2009, the year Bitcoin was launched.

4 Understanding the Mechanism of Blockchain

4.1 Foundations in Mathematics

Teaching applied mathematics is a huge challenge with respect to students' motivation. Basically, all the topics of business and financial mathematics were developed and published some hundred years ago. For example, calculus stems from the 17th century (Newton and Leibnitz), but is still a fundamental principle without which our civilisation would not function or would at least look very different.

Let us go back to the roots of blockchain to find out what mathematical topics provided the foundation for the blockchain mechanism. Its roots go back to the 17 and

18th century (Fermat, Euler), to the algebraic number theory which is the elementary foundation of cryptography (Hoffstein et al. 2008; Katz and Lindell 2008).

It is evident that it does not make much sense to teach blockchain at a business school by starting with basic maths, which is the fundamental fabric of blockchains. On the other hand, without an understanding of the principles of cryptographic hashing, public-private keys and the stochastic process of mining, it is not possible to recognize the basic principles of the blockchain mechanism; and without an understanding of the basic principles, it is not possible to evaluate or discuss different fundamental decisions of how and when blockchains could or should be used. The time might now have come to rethink the business and financial mathematics syllabus and spend some time on mathematical topics more related to business instruments of the 21st century.

Unfortunately, most current textbooks for business mathematics are the size of reference books with up to a thousand pages, e.g. Harshbarger (2009) or Haeussler et al. (2005). In a few years, these authors will probably add some twenty pages or more about the mathematical foundation of blockchain to their textbooks.

4.2   Foundations in Technology

Looking at the main components of the Bitcoin protocol, we recognize that the crypto-properties of blockchain are not sufficient to make Bitcoin become a working exchange platform. To understand Bitcoin, we must also discuss peer-to-peer networks, distributed consensus, shared ledgers, smart contracts and chain code, etc. Mougayar (2016, p.18) specifies ten properties which he regards as relevant for the understanding of blockchain:

9   Cryptocurrency
8   Computing Infrastructure
7   Transaction Platform
6   Decentralized Database
5   Distributed Accounting Ledger
4   Development Platform
3   Open Source Software
2   Financial Services Marketplace
1   Peer-to-Peer Network
0   Trust Services Layer

This classification is worth discussing although its structure does not seem completely logical to me. I would transfer layer 2 "Financial Services Marketplace" and layer 5 "Distributed Accounting Ledger" to the top, together with layer 9 "Cryptocurrency". Also, layer 3 "Open Source Software" should be renamed "Software" because a very relevant blockchain software called "Guardtime" is not open source (Williams-Grut 2016). The new order would then be:

9 Cryptocurrency
8 Financial Services Marketplace
7 Distributed Accounting Ledger
6 Computing Infrastructure
5 Transaction Platform
4 Decentralized Database
3 Development Platform
2 Software
1 1 Peer-to-Peer Network
0 Trust Services Layer

Real transactions in a blockchain could be shown using a Bitcoin browser. Ether or Hyperledger fabric could be chosen to provide demonstrations of smart contracts.

Another approach to explain the main structure of a blockchain was taken by Burgwinkel (2016, p. 5), who starts his explanation of the blockchain with the following properties:

- Blockchain technology
- Blockchain software
- Blockchain applications
- Blockchain platforms
- Blockchain-as-a-Service

We used this structure several times when teaching executives. Its strength lies in its briefness and in the fact that these layers follow a pragmatic approach which reflects a service layer architecture of IT infrastructure. The advantage of using existing terms and principles for blockchain come at a price. The main impact of blockchain by redefining the classical IT architecture is veiled (Iansiti and Lakhani 2017).

5 Applications of Blockchain

When considering relevant fields of application of blockchain, the number one topic is finance. This again is due to Bitcoin, which is perceived as a direct assault on the finance industry. However, how could blockchain be used beyond a cryptocurrency?

Contracts, transactions and records are the basic instruments for business processes or relations between people, companies, organizations or even nations. Until now, the internet has not able to guarantee the secure exchange and storage of business transactions, whereas blockchains promise to solve this problem (Iansiti and Lakhani 2017). Indeed, blockchain can do much more than deliver a value with tokens. Blockchains could be used in any industry to conduct business transactions.

For Bitcoin or finance related applications few textbooks can be found which deliver a curriculum for teaching. "Bitcoin, Blockchain und Kryptoassets" (Berentsen and Schär 2017) is a textbook written in German that delivers a very fundamental introduction to finance theory combined with Bitcoin technology.

Recently, besides finance, many further application fields have been substantiated (Huckle et al. 2016; Nachiappan et al. 2016; Burgwinkel 2016):

- Fintech
- Insurance
- Notary Public
- Government
- E-Health
- Energy Markets
- Internet of Things
- Supply Chain
- Transportation

There are interesting business cases which can help to explain the relevance of blockchain for these industries. Most of these are sponsored by organizations or companies which offer consulting or services, for instance, IBM (n.d.), Guardtime (n.d.) or associations with specific industries or public services GBA (n.d.). It might take a bit longer, however, until literature with a systematic analysis of blockchain use in different industries will be available.

6 Teaching Blockchain: The Experience of Knowledge Illusion

Teaching material about blockchain can be pitched at a very abstract level, which does not help define a blockchain syllabus. Tapscott and Tapscott (2017) extend the relevance of blockchain for higher education far beyond teaching but predict that the whole education system will be changed completely.

Teaching notes are more helpful for blockchain but, unfortunately, most of them are either not up-to-date (Watters 2016) or cover a specific aspect of blockchain which does not meet my scope of teaching. The majority of them focuses on blockchain and cryptocurrencies, which is related to Bitcoin as a showcase (Educase 2016; Berentsen and Schär 2017). Another cluster of material which partially overlaps with these is devoted to technical aspects and programming (Zeltsinger; Princeton University 2016).

I close with some remarks about my personal learning and teaching experience. I only started teaching blockchain after two years of studying different aspects of blockchain foundations. I was influenced by the wide variety of approaches to this topic, and I felt unsure about the right way to go about it. At that time I read a new publication about cognitive science entitled: "The knowledge illusion" (Sloman and Fernbach 2017). It fully reflected my perceptions of my own learning experience:

The knowledge illusion occurs because we live in a community of knowledge and we fail to distinguish the knowledge that is in our heads from the knowledge outside of it. We think the knowledge we have about how things work sits inside our skulls when in fact we're drawing a lot of it from the environment and from other people....Because we confuse the knowledge in our heads with the knowledge we have access to, we are largely unaware of how little we understand. We live with the belief that we understand more than we do....[M]any of society's most pressing problems stem from this illusion (Sloman and Fernbach 2017, pp. 127–129).

Having learned about blockchain by reading many books, attending several MOOCS and experimenting with Bitcoin, I had the illusion that I understood blockchain quite well. However, when I started to develop teaching material for my students, I realized that I could not explain how blockchain really works. I became aware that I had huge deficits in many aspects of blockchain technology and its application. I had to invest much more time and passion until I felt able to explain blockchain to a class.

As soon as I realized this knowledge illusion, I paid more attention to my environment. Not surprisingly, I could confirm Sloman & Fernbach's thesis with my experience from discussions or lectures about blockchain with students, business people and other academics. As soon as people have learned at least a few facts about blockchain, they feel much more confident in their understanding of the topic. Inspired by experiments about cognitive reasoning, I carried out a survey in class about blockchain. I worked with six students over a period of several weeks to learn about a range of blockchain topics and prepare introductory presentations to their class. We then presented our learning to the class in a one-day seminar.

At the beginning of this day, I asked the whole class how they assessed their knowledge in this topic. At the end of the day, I asked the same question again. The students' answers are shown in Fig. 4. From a teacher's point of view, this result is very positive. The students appreciated what they had learned, which was also confirmed by the qualitative feedback afterwards. From an expert's point of view, the results are questionable. In fact, the students had received a superficial introduction to blockchains, which was followed by short presentations of three different practical applications. This was far removed from what I would call a reasonable knowledge level in any field of blockchain. Indeed, such a complex topic cannot be conveyed to novices in a one-day seminar.

7 Conclusion

Defining a syllabus to teach blockchain is more like going on a journey than delivering a list of topics. This is due to the fact that blockchain development is still at an early stage. The internet provides a countless number of short instructions on Bitcoin or blockchain, which do not help to gain a real understanding of the topic. The revision of more substantial teaching material shows a fragmented picture, and depends on the

What is your expertise in Blockchain?

Before Blockchain Seminar		N		
1	not existing	4	13%	
2	quite small	19	59%	
3	medium	8	25%	
4	excellent	1	3%	
	Total	32	100%	

After the Blockchain Seminar		N		
1	not existing	1	4%	
2	quite small	9	32%	
3	medium	17	61%	
4	excellent	1	4%	
	Total	28	100%	

Fig. 4 Self-assessment of students before and after a one-day introductory seminar on blockchain

author's initial approach and on the focus. Figure 5 should be read as a structured list of relevant topics about blockchain in the light of three main approaches: Business impact, Building blocks and Application of blockchain. This list is a snapshot and it will change quite quickly. It is also important to keep the whole picture in mind when focusing on a subject because the decomposition of blockchain into its elements would destroy its main nature.

The following questions about blockchain teaching could not be answered in this study and might deserve additional research:

- How can students become more involved in the sense of a self-activated learning experience? What would be an appropriate form for a "blockchain-lab" for business students?
- How can we design a teaching environment and mentally involve more faculty with diverse professional and scientific backgrounds as the motor for an ongoing learning community on this topic?
- What indicators can be used to define the relevant trends in blockchain to find a path between messing about and lagging?

Fig. 5 Blockchain topic
structure for teaching
business students, snapshot
2017

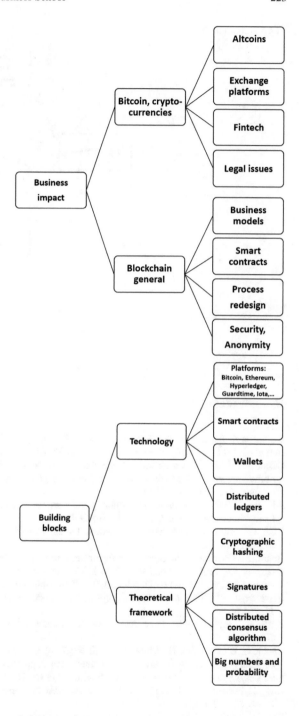

Fig. 5 (continued)

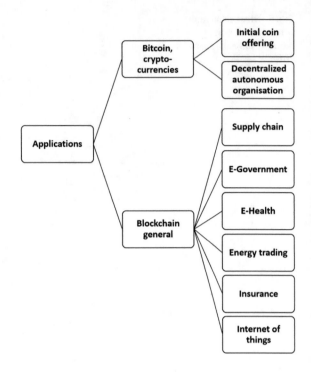

References

Berentsen A, Schär F (2017) Blockchain Buch—Bitcoin, Blockchain und Kryptoassets

Burgwinkel D (ed) (2016) Blockchain technology. De Gruyter, Berlin, Boston

Castor A (2017) Consensus 2017: Even Academics can't keep pace with Blockchain change. http://www.coindesk.com/consensus-2017-even-academics-cant-keep-pace-with-blockchain-change/. Accessed 1 Jun 2017

Educase (2016) Blockchain. https://library.educause.edu/resources/2016/9/7-things-you-should-know-about-blockchain. Accessed 16 Mar 2017

GBA Government Blockchain Association. https://www.governmentblockchain.org/resources/. Accessed 13 Jun 2017

Guardtime Guardtime Solutions. https://guardtime.com. Accessed 13 Jun 2017

Haeussler EF, Paul RS, Wood RJ (2005) Introductory mathematical analysis for business, economics, and the life and social sciences, 13, 2014th edn. Pearson Education

Harshbarger RJ (2009) Mathematical applications for the management, life, and social sciences, 11, 2015th edn

Hoffstein J, Pipher J, Silvermann JH (2008) An introduction to cryptography. Springer, New York, NY

Huckle S, Bhattacharya R, White M, Beloff N (2016) Internet of things, Blockchain and shared economy applications. In: Procedia computer science, pp 461–466

Iansiti M, Lakhani KR (2017) The truth about Blockchain. HBR 118–127

IBM Hyperledger and IBM Blockchain. https://www.ibm.com/blockchain/hyperledger.html. Accessed 13 Jun 2017

Katz J, Lindell Y (2008) Introduction to modern cryptography. Chapman & Hall/CRC

Mougayar W (2016) The business blockchain: promise, practice, and application of the next Internet technology. John Wiley & Sons, Inc., Hoboken, New Jersey

Nachiappan PP, Crosby M, Pattanayak P, Verma S, Kalyanaraman V (2016) BlockChain technology: beyond Bitcoin

Nakamoto S (2008) Bitcoin: a peer-to-peer electronic cash system, pp 1–9

Princeton University (2016) Bitcoin and Cryptocurrency Technologies. https://www.coursera.org/learn/cryptocurrency/home/welcome

Sloman S, Fernbach P (2017) Knowledge illusion why we never think alone. Riverhead Books, London

Tapscott D, Tapscott A (2016) Blockchain revolution: how the technology behind bitcoin is changing money, business, and the world

Tapscott D, Tapscott A (2017) The blockchain revolution and higher education I EDUCAUSE. In: Education review. http://er.educause.edu/articles/2017/3/the-blockchain-revolution-and-higher-education. Accessed 1 Jun 2017

Watters A (2016) The blockchain for education: an introduction. http://hackeducation.com/2016/04/07/blockchain-education-guide. Accessed 1 Jun 2017

Williams-Grut O (2016) Estonia is using the technology behind bitcoin to secure 1 million health records. Bus Insid

Yli-Huumo J, Ko D, Choi S, Park S, Smolander K (2016) Where is current research on Blockchain technology? - A systematic review. PLoS ONE 11(10): e0163477. https://doi.org/10.1371/journal.pone.0163477

Zeltsinger S What to teach when teaching blockchain - Shlomi Zeltsinger. https://zeltsinger.com/2017/04/07/teach-teaching-blockchain/. Accessed 1 Jun 2017

Computational Intelligence in Modelling, Simulation, Optimization, and Control

Thomas Hanne and Rolf Dornberger

Abstract One of the biggest trends nowadays is to make IT systems more intelligent in order to solve problems which were previously too complex to be solved, or where the computing power prevented solving them within reasonable time. This trend is the renaissance of Artificial Intelligence (AI). In this chapter, we first discuss some general issues related to mathematical modelling, simulation and optimization for practical applications. After this, we review certain related techniques referred to as Computational Intelligence (CI), which are particularly useful for dealing with complex mathematical models. We discuss the foundations of CI techniques and relations to AI. In the third section, we discuss selected areas of application of CI and AI for business improvement on a more detailed level, i.e. in transportation planning, in warehouse management, and in robotics, in order to motivate their modeling complexity and potential for CI applications. We conclude with an outlook of CI techniques concerning their expected forthcoming practical importance.

Keywords Computational intelligence · Artificial intelligence · Applications Modeling · Simulation · Control

1 The Role of Mathematical Modeling, Simulation, and Optimization

It has often been mentioned that mathematics began to play an increasing role in the field of economics, especially during the 20th century (see, e.g. Boulding 1948).

T. Hanne (✉)
Institute for Information Systems, University of Applied Sciences and Arts Northwestern Switzerland, Riggenbachstrasse 16, 4600 Olten, Switzerland
e-mail: thomas.hanne@fhnw.ch

R. Dornberger (✉)
Institute for Information Systems, University of Applied Sciences and Arts Northwestern Switzerland, Peter Merian-Strasse 86, 4002 Basel, Switzerland
e-mail: rolf.dornberger@fhnw.ch

© Springer International Publishing AG 2018　　　　　　　　　　　　　　　227
R. Dornberger (ed.), *Business Information Systems and Technology 4.0*,
Studies in Systems, Decision and Control 141,
https://doi.org/10.1007/978-3-319-74322-6_15

This was first observed in the formalization of economics by neoclassic models or in Keynesian approaches. Stronger management-oriented approaches such as in production, finance and investment, or logistics saw a related "mathematization" especially from the 1950s onwards when operations research emerged as a new discipline for solving practical business problems.

In operations research, the main goal was no longer to describe economic variables in a mathematical form. Instead, its main focus was decision oriented, where mathematical models were expected to lead to better solutions for real-world problems occurring in companies or other institutions. Besides the usage of statistics, simulation and optimization (sometimes also referred to as mathematical programming) became major subareas of research in the new discipline. Before the 1950s, mathematical models were developed and used occasionally for problems such as calculating economic order quantities or solving specific problems in transport planning. But now, mathematics seemed to promise a universal applicability to problems of all sorts. One of the milestones in this development was the introduction of the simplex algorithm by Dantzig, which was able to solve linear optimization problems as they appeared, for instance, in the field of production planning (Dantzig 1982).

Already in the early days, there was a great deal of discussion concerning the usefulness of the "mathematization" of economics and related fields. On the one hand, the adequacy of the economic models concerned, their efficiency and effectiveness, or possible insights from studying them were questioned. This discussion was less intense for the more practically oriented models, but here, too, the model assumptions were challenged or the availability and certainty of required data were doubted. On the other hand, there were obvious success stories despite the limitations of the mathematical models used (Hanne and Dornberger 2017). These limitations led to a strong impetus to further develop and refine the models concerned in order to make them suitable for practical use.

Some of the main challenges for the early approaches in operations research are as follows:

1. The assumed type of model (e.g. linear optimization model) does not reflect the real situation. This aspect is mainly due to the fact that in the early days a strong emphasis was placed on linear optimization problems, which assume continuous (real-valued) variables, whereas often in reality variables can be assumed to be integer (or discrete in some other way). For instance, quantities of goods are often measured in pieces so that non-integer solutions do not appear to be feasible. Although simple loopholes (such as the rounding of solutions) may be appropriate in some cases, there are situations where this might lead to completely unsatisfactory solutions. In particular, when the values are small or serve to represent combinatorial solutions (e.g. sequences of items) rounding may lead to very poor or infeasible solutions. This fact was discovered quite early and various approaches such as branch and bound or cutting plane methods were suggested, but their performance, especially for large-scale problems, often appeared to be a major drawback (see point 3).

2. Even if the basic model type (and, in particular, the domain of variables) appears to be basically suitable, the models turn out to be too simple to represent a real-life situation. There are two main reasons for this: On the one hand, in reality there are often further aspects to be considered which define whether a solution appears to be feasible or not. This means that further constraints related to these aspects should be considered in the problem formulation. Some of these constraints should possibly not be considered as "hard" but rather as an expression of something which is not desirable, but could be accepted if there is no way to avoid it, or if a solution is otherwise superior to others. On the other hand, such further aspects could be considered as additional objectives or criteria for the decision making process. Thus, such a situation could be considered by defining a multi-criteria or multi-objective problem, an area of research that was developed mainly from the beginning of the 1970s.

3. Since the development of complexity theory, many practically relevant optimization problems (especially discrete or combinatorial problems) have been characterized as non-deterministic polynomial-time hard (NP-hard). This theoretical property is the basis for assuming fundamental limitations in solving them to optimality (due to the $P \neq NP$ conjecture) because the runtime of any known algorithm increases exponentially (under worst-case conditions) with the problem size. As a consequence of this insight, but also due to practical considerations (such as acceptable runtimes of algorithms), the research goals have shifted somewhat from the desire for optimality towards the pursuit of algorithms delivering fast and robust solutions.

4. Last, but not least, it should be mentioned that it is often difficult to specify all required data (such as problem descriptions) completely and beyond doubt. On the one hand, it is well known that a company often cannot provide all data such as, for instance, available inventory, lead times in procurement, required resources and time to manufacture, or dependencies due to the bills of materials (e.g. because these data were never captured systematically). On the other hand, at least some of the data is affected by vagueness or uncertainty. For instance, for many companies future demand for their products appears to be uncertain. There might be appropriate approaches to forecasting, but even then, exact predictions cannot be expected. A simple approach is to use only roughly determined prediction values in the planning models concerned (e.g. for production planning). Since the early days of operations research, however, the focus has also often been on considering stochastic models, which assume that some or all of the problem data are random variables. It is often difficult to specify such models in more detail (such as assuming specific distributions of the random variables), or else the underlying phenomena are not really uncertainties but rather incomplete knowledge. Because of this, some further approaches were worked out to deal with difficulties in this regard, for instance fuzzy set theory.

As a consequence of these challenges, various novel methods have been developed and, in particular, the area of Computational Intelligence (CI) has emerged as a set of approaches to address them.

2 Computational Intelligence and Artificial Intelligence: Similarities and Differences

The term "Artificial Intelligence" (AI) is about 70 years old, whereas "Computational Intelligence" dates back about 25 years. Additionally, two different definitions are often provided for CI where, firstly, CI comprises a set of nature-inspired algorithms and secondly CI is understood as the standard approach to using computers as the basis for Artificial Intelligence. In the next subsections, we briefly present the history of AI and its rebirth as well as the emergence of CI.

2.1 The Renaissance of Artificial Intelligence

Nowadays, with increasing computational power, Artificial Intelligence is experiencing its renaissance. In some branches of AI, such as artificial neural networks (ANN or simply NN), the usage of very large NN systems is now making it possible to compute satisfying results in a reasonable time period. A wonderful example is speech recognition where deep learning (Schmidhuber 2014)—a particular extension of NN—delivers very promising results.

While AI was "scientifically born" in the 1950s (McCarthy 1955) and recognized as the basis for making machines truly intelligent, it provided two main streams in the following years: Strong AI and weak AI. In strong AI, the creation of human-like machine intelligence (in the form of software agents, perhaps enriched by a hardware robot) was the aim. Here, the intention was to develop human cognition, including consciousness, emotion and creativity on a human level or even better. In weak AI, research focused more on the implementation of human-like approaches to software such as reasoning, learning, adaptation, classification, etc. In the last century, due to limited computational power, weak AI was far more successful compared to strong AI and led, for example, to probabilistic methods in reasoning, classifiers and statistical learning methods, search and optimization methods, neural networks, symbolic-based logic and control theory, and programming languages for AI. However, today, strong AI is back—intelligent machines increasingly embodied in hardware robots (Pfeifer 2007) are flooding academia, business and society.

2.2 The Emergence of Computational Intelligence

A research and application area strongly related to AI is Computational Intelligence (Hanne and Dornberger 2017). In principle, CI translates well-known techniques from nature into computer algorithms. For instance, nature provides well-proven techniques such as evolution or swarm behavior in order to cope with the real-world problems of surviving and being superior to other species over millions of years. The discipline of CI collects these methods under the umbrella of "nature-inspired" and now comprises the main branches of Evolutionary Computation (EC), Swarm Intelligence (SI), Neural Networks (NN), Fuzzy Logic (FL), Artificial Immune Systems (AIS) and other related classes of nature-inspired methods.

Particular interest in CI started in the 1990s (Poole et al. 1998). In contrast to AI, CI focusses on sub-symbolic techniques while implementing nature-inspired methods in computer algorithms. Consequently, by using such heuristic, non-deterministic methods, CI is able to approximate and/or converge to solutions even in cases, where the search space is changing or the targets are moving (e.g. in control problems). Further advantages of CI methods are that it is sufficient to provide rough solution models, because CI methods are fault-tolerant to intermediate bad candidate solutions, and that they can mostly be parallelized very well. Nevertheless, the biggest disadvantage of CI methods is the often missing mathematical proof of their performance.

Similar techniques, more or less closely related to AI and CI, are Machine Learning (Mitchell 1997), Soft Computing, Cognitive Computing, Natural Computing (NC) and Bionics respectively Bionic Engineering (BE). While the former three techniques comprise purely computational methods, NC also focusses on the rebuilding of natural habits into computer hardware, and BE deals mainly with the adoption of traits in nature for technical products (Rechenberg 1973).

2.3   Branches of Computational Intelligence

CI methods are characterized by the expression "nature-inspired" and are sometimes already mentioned in AI. However, nature-inspired might partly describe very different mechanisms. Additionally, due to its "nature-inspired" character the wording generally used to describe these algorithms adopts words from biology.

One of the most important mechanisms is the evolution found in Evolutionary Computation (EC) and often in Swarm Intelligence (SI):

- EC addresses the evolution by using the principle of survival of the fittest. In order to find increasingly better solutions in the optimization or search problem, a set of candidate solutions (the so-called "individuals") are improved over several iterations (named "generations") using particular operators denoted as "selection", "crossover" and "mutation".
- SI also improves the quality of its candidate solutions (so-called "agents"), but by using principles of collective behavior and self-organization. These agents interact locally, but produce intelligence-like global behavior. Applications are often in optimization and the search for problems for distance or time minimization.

Further branches of CI, adapting other nature-inspired principles, are briefly described as follows:

- Neural Networks (NN) are another branch of CI methods. They use the principles of learning mechanisms in biological brains consisting of interconnected neurons.

The NN are usually trained to match certain input data to particular output data, while values are processed through a sequence of hidden layers performing simple calculation operations in each neuron.

- Fuzzy Logic (FL) extends the binary arithmetic of computers to a many-valued logic. It permits the approximation of solutions or conclusions with incomplete or ambiguous data while introducing the principle of fuzzy human reasoning—to a certain degree.
- Artificial Immune Systems (AIS) apply the principles of the immune systems of living creatures to detect entering viruses (medically speaking "pathogens"). Here, processes of learning and remembering patterns are used for problem solving in adaptive or matching systems.
- Other methods are also counted as CI, such as Reinforcement Learning or Simulated Annealing, where the latter is from non-living nature, but related to thermodynamic principles in physics. Furthermore, different kinds of heuristics (so-called rules of thumb), metaheuristics (heuristics based on general concepts such as those found in nature), and hyperheuristics or hybrid heuristics (combinations of different (meta)heuristics mostly in the context that one heuristic is "controlling" another heuristic) are partly taken into account as CI methods or at least as an add-on of CI-methods: Tabu Search, Neighborhood Search, Memetic Computation, Differential Evolution, Learning Classifiers etc. In this terminology, evolutionary algorithms and swarm algorithms are also known as metaheuristics.

2.4 The Big Picture of Computational Intelligence

Figure 1 presents the big picture of CI methods, focusing on the sub-methods of EC and SI (see Hanne and Dornberger 2017, p. 19): CI comprises many branches, i.e. Evolutionary Computation, Swarm Intelligence, Neural Networks, Fuzzy Logic and further related nature-inspired methods such as Reinforcement Learning. These branches of CI comprise further methods (in this figure only a focus on EC and SI is provided, where Swarm Algorithms belong to both classes). Parallel to CI, Artificial Intelligence exists, as do further methodologies such as Bionics, Soft Computing, Natural Computing, and Machine Learning, which partly comprise the same or similar methods.

CI methods count as heuristics and metaheuristics. They are used to find (near) optimal solutions although this optimality can (often) not be proven except by checking all possible solutions. CI methods are often applied when real-world problems, respectively their hard-to-solve computational models, have to be optimized. CI methods are (mostly) non-deterministic, thus, they follow a particular stochastic convergence path within every run.

CI methods are used in single as well as in multi-objective optimization. The problem of simply optimizing variables is often turned into a learning or classification problem, e.g. pattern or speech recognition. CI methods are also used to design controllers or technical processes, to simulate biological evolution or brain functionality,

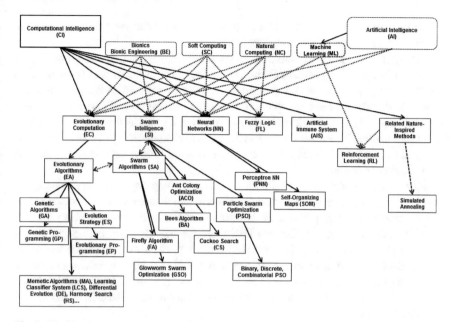

Fig. 1 The big picture of computational intelligence: CI comprises many branches, i.e. evolutionary computation, swarm intelligence, neural networks, fuzzy logic and further related nature-inspired methods such as reinforcement learning

or to create music or art. However, one of the fastest growing areas of application for CI and/or AI nowadays is the field of robotics.

3 Selected Application Problems for Computational Intelligence Approaches

In this section, we discuss typical problems from three selected areas of application which have already benefited significantly from the application of CI techniques: transportation, warehousing, and robotics. Further areas related to logistics and supply chain management are discussed in Hanne and Dornberger (2017).

3.1 Problems in Transportation

Transportation problems are among the most relevant planning tasks in logistics. The overall transportation market was estimated to have a volume of US\$ 4,700 billion in 2016 which corresponded to 6% of the global gross domestic product (Plunkett Research 2017). Therefore, an optimization of the respective transport processes and related cost savings can be considered as key drivers for further growth, increase in welfare, and the continuation of globalization.

Transport planning tasks are among the oldest and most frequently considered problems in operations research. The classic transport problem was formulated as early as 1941 and was related to military activities of the allied forces (Gass and Assad 2005). The original transport problem is a linear optimization problem, which assumes that demand at different locations is given as well as the respective supply quantities at other locations. The task is then to find an assignment of source locations to demand locations in such a way that their demand is covered and total transport costs become minimal.

However, in many practical situations, transport planning duties are completely different. Often it is clear from which location to which other location transportation should take place, e.g. when there is a unique warehouse from which to cover a specific customer demand. Besides this, there are usually other aspects that have to be considered in a more differentiated form, e.g. details of the actual transport. In particular, it is often necessary or useful to combine several deliveries in a single transport (e.g. making several deliveries simultaneously with the same vehicle). Here, the specific sequence of locations within a tour determines its length, duration, and associated costs.

The simplest variant of such tour building problems is the Travelling Salesman Problem (TSP), which assumes that there is a given number of locations to be visited during a single tour that starts and ends at a defined depot. The goal is thus to find a sequence that includes all locations (except the depot) exactly once (also referred to as a permutation) which results in the lowest costs.

Despite its simplicity in formulation and understanding, the TSP is NP-hard. Consequently, a large amount of research has addressed the problem and especially heuristics and metaheuristics have been worked out to solve the problem. Simple heuristics can often be characterized as a form of local search. This means that from a given solution improvement steps are considered which explore other solutions in the "neighborhood" of the current solution or which are similar to the current solution. For instance, minor changes to a given solution could be considered, such as the reinsertion of a location at a different position within the sequence, the exchange of two locations or the reversal of a subsequence within a given sequence.

CI methods and other metaheuristics are often based on such basic concepts for improvement within a more complex framework but follow some further search principles such as, for instance, natural evolution. Genetic algorithms may use such "local changes" as mutation steps and combine them with a recombination of solution parts from different current solutions (crossover).

Another general requirement for the application of metaheuristics is the representation of solutions within the algorithm. A straightforward representation is to represent a sequence as an array of integer numbers which correspond to the location IDs (usually the numbers $1, \ldots, n$ representing the n locations to be visited during a tour and possibly using "0" to represent the depot location). This direct representation of a tour is easy to understand but has some drawbacks. Most notably, the generation of new solutions during the run of the metaheuristics may lead to infeasible solutions. For instance, if a feasible sequence is used (with each number occurring exactly once) and only one entry in it is changed, then it will always result in an infeasible solution because the new entry can already be found at another position in the array. In addition, standard approaches for crossover in genetic algorithms would usually lead to infeasible solutions as some locations may appear more than once while others were missing in the array.

Therefore, adapted change mechanisms (for variation operators) for metaheuristics are reasonable when they consider the feasibility of new resulting solutions. With respect to evolutionary algorithms, a considerable number of suitable mutation operations and crossover operations have been suggested in the literature (for more details see Hanne and Dornberger 2017) and similar approaches can be applied for other metaheuristics.

Another possibility is, however, to use robust solution representations, which are not so vulnerable due to solution variations. One of the most frequently used alternative representations to make use of such an advantage is random key encoding. Here, the sequences are not directly represented but instead, for each location a kind of priority value is stored which defines the position of the respective location within a sequence. Thus, solutions are represented by an array of real-values. Given such an array, the actual sequence can be derived by sorting the locations according to the assigned real values. Such a representation, in particular, ensures that for any given solution representation, the corresponding sequence always includes each location once. Moreover, standard variation operators for solutions represented by arrays of real numbers can be applied without any modification.

Although the above-mentioned TSP is probably the best-known example of sequencing and routing problems, it is usually insufficient from a practical point of view in transport planning. Many real-life aspects such as constraints relating to the capacities of vehicles, the feasible visiting times of locations, or the overall lengths of tours are neglected. Such aspects are considered in further mathematical problem formulations.

The simplest and most important generalization of the TSP is the Capacitated Vehicle Routing Problem (CVRP), which introduces a capacity constraint: Every time during each tour, the total load of carried goods in a vehicle may not be larger than a specified capacity limit. As a consequence of this, it is no longer possible to execute all deliveries during a single tour and thus several tours are required. This means that either a considered vehicle can execute only a limited number of orders during a tour before returning to the depot to pick up further transport goods, or the transports must be carried out by several vehicles operating in parallel. In any case, during each single tour, the capacity requirements have to be strictly observed.

This problem can be considered a two-stage problem where the first stage defines the assignment of a transport order to a tour while the second stage requires finding good sequences for the orders assigned to each tour. Some approaches make use of this two-stage structure of the planning problem (possibly using different approaches for each stage) while others still try to solve the two problem stages in a single step. In either case, the metaheuristics used must consider the new problem formulation and variation operations may become more differentiated, for instance including transfers of locations or subsequences from one tour to another.

A broader survey of using metaheuristics for different kinds of transport and route construction problems can be found in Hanne and Dornberger (2017). Moreover, the variety of CI techniques and other metaheuristics considered with regard to such problems and the related results is discussed there in more detail.

3.2 Problems in Warehouse Management

Warehousing is usually considered the second most important activity in logistics. For instance, according to the State of Logistics Report (Kearney 2016) in the US, the inventory carrying costs in 2015 were US$ 427.3 billion, which corresponds to 2.37% of the GDP. Although this percentage was occasionally much higher in the past, there is still potential for further improvements and cost savings.

From a mathematical modelling point of view, different questions related to warehouse management can be asked. The most obvious one concerns the inventory quantities to be stored. The relating model in its simplest form considers the economic order quantities (EOQ). In this model, the warehousing costs are considered in combination with the ordering costs and a cost-minimal order quantity can be identified. Ordering means here that the required materials can either be procured externally (i.e. requiring transportation costs) or that they are produced internally in the form of lots. In both cases, it is assumed—in the simplest type of model—that the order costs are fixed for each order whereas the storage costs are assumed to be proportional to the respective stored quantities. The cost-optimal solution of this type of model, which can be expressed in equation form, dates back to Harris (1913).

Even slightly more complicated versions of the EOQ model no longer allow a similar expression for the optimal solution to be found and are already NP-hard problems (see e.g. Karimi et al. 2003) so that CI techniques are promising approaches for solving them. This concerns in particular various variants of capacitated lot sizing problems.

For instance, the single-item capacitated lot-sizing problem (SICLSP) is a multi-period model to determine the production and inventory quantities of a single material in order to minimize costs, which consist of variable production costs, setup costs (if production in the respective period takes place), and inventory costs. During each period, a given demand has to be covered by production and available inventory and a capacity limit for the production has to be observed.

In the multi-item version of this problem (MICLSP) the production and inventory of several products are to be planned. It is assumed that the production of all materials requires the same resources. Therefore, its capacity limit needs to be matched with the capacity consumption of all produced materials.

Another variant of lot sizing problems considers the production structure of materials as expressed in the bills of materials (BOMs). This means that not only are the production and inventory of end products to be planned but also predecessor parts on one or several levels. The total demand of a material then consists of primary or external demand and secondary or internal demand, which results when a material is required in order to produce a successor material (according to the BOM of the successor material). Even in its uncapacitated version, the multi-level lot-sizing problem (MLLSP) is NP-hard. Of course, this also holds for the capacity-constrained version of the problem, which is also referred to as a general capacitated lot-sizing problem (GCLSP).

Due to the hardness of the lot-sizing problems, approaches from the area of Computational Intelligence have frequently been considered in order to solve them, along with more simple heuristics and mathematical programming techniques. Similarly, with regard to vehicle routing problems, it can be said that the whole range of metaheuristics has been explored with respect to its applicability. There are, however, major differences in the model formulations compared to the above vehicle routing problems: Here, in particular, the decision variables are partially real values and partially discrete. For instance, in the case of the SICLSP and MICLSP decision variables such as production quantities are assumed to be real values (but could also be integer values) whereas additional dependent binary decision variables are used to determine whether production takes place or not in a given period. This may possibly be a major reason why evolutionary algorithms seem to dominate among the metaheuristics used to address such problems. Unlike some other metaheuristics, evolutionary algorithms are especially suitable to deal with both real-valued and binary variable domains. In fact, their main roots, genetic algorithms and evolutionary strategies, have been developed for binary and floating-point solution representations respectively, so that strategies to take into account the respective peculiarities are well established. Various examples of applying evolutionary algorithms and other metaheuristics to lot-sizing problems are discussed in Hanne and Dornberger (2017).

Since lot-sizing models only deal with the question of inventory levels (and this also just in a rough form since, for instance, aspects such as safety stocks are not included), further inventory-related questions remain open: Where (at which location(s)) should the inventory be stored? How should subsequent processes such as order picking be organized?

The first question can be formulated in different ways: From a strategic level, it could refer to questions of where, for instance, warehouses of a company should be located or how to (re-)design the supply chain. For instance, facility location problems and other location problems belong in this category (see Hanne and Dornberger 2017, Chap. 6). From an operational point of view, it is assumed that locations are already "given". Here, loosely speaking, problems with a focus on in-house operations can be distinguished from problems involving several warehouse locations.

In the first case, it is usually assumed that inventory within a given warehouse is to be managed, and this consists of a larger amount of storage locations like rack spaces corresponding to a pallet or a bin. For instance, for newly arriving inventory it is necessary to determine suitable empty storage locations. For the optimization of subsequent operations (especially the throughput in order picking) it makes sense not to assign possible storage locations according to a fixed plan or in some arbitrary way but to consider aspects such as demand frequency and volume requirements.

Another optimization problem occurs when materials are retrieved from their storage locations (e.g. by automated storage and retrieval machines (ASRM)) during order picking. Here a number of aspects offer decision alternatives and could be considered in different types of mathematical models: First of all, if several items (or transport units) can be transported together, the sequence of their pickings usually determines the transport distance and the time required. In fact, this problem corresponds to the CVRP discussed above which is usually interpreted in the context of longer distance transportations (truck transport). Besides such pure routing aspects, in-house transportation may also involve the planning of specific paths (if several ways are possible) considering objectives such as shortest transport times or a load balancing over different sections of the journey.

Further optimization potential can be provided by having options in choosing where to pick the items, i.e. when the required items are not only available at a single storage location. Apart from minimizing picking time, this may also allow for a better balancing of the workload of involved resources such as ASRM or human pickers.

If we consider inventory locations from a broader perspective which involves various possible locations such as warehouses, the main operational questions can be as follows: Which demand should be covered from which warehouse? When and how much inventory should we transport from which location within our supply chain to which other location? The first question can be considered as a variant of the above vehicle routing problems with several depots. The second question appears to be more complicated and is usually addressed in models denoted as inventory routing problems (see Sect. 4.7 of Hanne and Dornberger 2017). For instance, in the single link shipping problem, only two locations within a supply chain are considered. The demand of multiple items at a destination node is to be covered during all periods of the planning horizon. These goods are produced at the source node under a capacity constraint. Shipments from the source node to the destination node are also planned under consideration of limited capacities. Both locations can keep inventory but with possibly different holding costs. The overall objective is the minimization of costs which consist of inventory costs (at both locations) and transport costs which are assumed to be a fixed amount which occurs when transportation takes place within a period.

Similar to lot-sizing problems, these problems are usually computationally hard (NP-hard) and involve real-valued and discrete variables. For this reason, computational intelligence approaches appear to be a very attractive way to solve them. Compared to other warehouse management problems, however, there are less CI based studies so far, which may be due to the more difficult nature of inventory routing problems and the fact that it is a less explored area of optimization problems in general.

3.3 Problems in Robotics

Robotics is the interdisciplinary field of combining moving hardware with (more or less intelligent) software applications. While this combination was already predicted in strong AI in the last century, not much progress was made except in sci-fi movies. Today, there is a great deal of hype in the research, engineering and application of robots. Thus, the new discipline of robotics includes the design, construction, operation and maintenance of robots (Siciliano and Khatib 2017).

As humans are increasingly tending to design robots with human-like appearance and behavior, it seems likely that nature-inspired methods might be preferred to design and control robots. Besides AI, the different branches of CI are used nowadays for designing or controlling robots. In the following, some examples of the application of CI and/or AI are given:

- EC can be used to optimize the mechanical design of the robot simulating the evolution process in nature, which has led to today's living species on earth. EC permits suitable configurations of a particular robot used in a specific environment for specific tasks to be found (Pfeifer 2007): The entire configuration is optimized, e.g. the number and size of the wheels of driving robots, or the dimensions of the legs (thigh and lower leg) of walking robots. In addition, the position and the rotation angles of electricity-powered mechanical joints can be organized and optimized.
- The activation of robots in sophisticated (small-scale) movements, mostly by the synchronization of a set of electro motors, is a control problem. The controllers can classically be designed using control theory. Nevertheless, as the robots are generally exposed to a non-perfect real-world environment, learning algorithms are suitable for designing robot controllers. Locomotion, i.e. the guided (large-scale) movement of a robot, and manipulation, the ability to perform mechanical tasks, are also specific control problems (Siciliano et al. 2009). Thus, EC, NN and FL are very often used.
- Furthermore, the robot should be provided with a certain degree of intelligence. This (robot) intelligence is mostly a set of specific intelligence-like features such as a sense for the surrounding. Thus, computer vision allows robots to detect and identify their environment. Further intelligent-like features could be hearing, touching, smelling and so on. Thus, CI/AI are suited to make the robot intelligent.
- A particular kind of intelligence is navigation and interaction with the environment, known as autonomous navigation, where CI/AI methods support the robot.
- Furthermore, the interaction with other robots and with humans is the ultimate goal. Here, the recognition of speech, face and gestures is just the beginning. The development of artificial emotions, social intelligence and personality is still in its infancy.

CI and AI will play a tremendous role in the development and control of such characteristics of robots in the future.

4 Conclusions

In practice, complex planning problems in business and society such as those discussed above represent major challenges for which adequate solutions must be found. The main reason is assumed to be their computational hardness, which clearly limits the possibility of implementing efficient algorithms. Approaches from the field of CI have been developed to deal with such difficulties: They can usually provide fast solutions, which appear to be good but not optimal. The approaches and calculated solutions are frequently robust so that they also appear to be reasonable, when certain problem parameters change or are not completely known in the beginning. The approaches are flexible so that they can be applied with a moderate effort to adapt to a wide range of problem types.

The recent past has seen a considerable extension of the field of CI. In particular, many new metaheuristics, especially particular branches of CI such as swarm algorithms, have been developed during the last ten years. Despite the rapid growth of the research field, it has not been possible so far to explore all the new algorithms with regard to their suitability for the problem types discussed in this chapter. It is expected that research will address these aspects in the years ahead, which will also see the creation of further CI methods and variations and improvements of existing approaches.

Moreover, it is expected that not only will academic research address the problems discussed but that also the actual usage of such approaches in practice will increase significantly. In order to achieve this, it will become essential to provide the relevant algorithms by means of software designed for easy use and integration into existing information systems such as enterprise resource planning (ERP) software, which is widely used by companies.

References

Boulding KE (1948) Samuelson's foundations: the role of mathematics in economics. J Polit Econ 56(3):187–199

Dantzig GB (1982) Reminiscences about the origins of linear programming. Oper Res Lett 1(2):43–48

Gass SI, Assad AA (2005) An annotated timeline of operations research: an informal history, vol 75. Springer Science & Business Media

Hanne T, Dornberger R (2017) Computational intelligence in logistics and supply chain management. Springer, New York

Harris FW (1913) How many parts to make at once. Fact Mag Manag 10(2):135–136

Karimi B, Ghomi SF, Wilson JM (2003) The capacitated lot sizing problem: a review of models and algorithms. Omega 31(5):365–378

Kearney AT (2016) 27th annual state of logistics report. The Council of Supply Chain Management Professionals, Chicago

McCarthy J, Minsky M, Rochester N, Shannon C (1955) A proposal for the Dartmouth Summer Research Project on artificial intelligence. Archived from the original on http://www-formal. stanford.edu/jmc/history/dartmouth/dartmouth.html. Accessed 30 Nov 2013

Mitchell TM (1997) Machine learning. McGraw-Hill

Pfeifer R (2007) How the body shapes the way we think: a new view of intelligence. The MIT Press

Plunkett Research (2017) Transportation industry statistics and market size overview, business and industry statistics. https://www.plunkettresearch.com/statistics/Industry-Statistics-Transportation-Industry-Statistics-and-Market-Size-Overview/. Accessed 4 Mar 2017

Poole DL, Mackworth AK, Goebel R (1998) Computational intelligence: a logical approach. Oxford University Press

Rechenberg I (1973) Evolutionsstrategie—Optimierung technischer Systeme nach Prinzipien der biologischen Evolution. Frommann-Holzboog, Stuttgart

Schmidhuber J (2014) Deep learning in neural networks: an overview. Neural Netw 61:85–117

Siciliano B, Khatib O (2017) Springer handbook of robotics, 2nd edn. Springer

Siciliano B, Sciavicco L, Villani L, Oriolo G (2009) Robotics: modelling, planning and control. Springer

Innovation Potential for Human Computer Interaction Domains in the Digital Enterprise

Stephan Jüngling, Jonas Lutz, Safak Korkut and Janine Jäger

Abstract This chapter summarizes a historic overview of some iconic examples of human computer interaction devices and focuses on a human computer interaction paradigm which is based more on human language. Human language is by far the most utilized means of conscious communication between humans whereas the mouse and keyboard are the dominant means to store and process information in computers. This chapter elaborates on the main challenges related to human language, as well as on ideas showing how human language, written or spoken, is embedded in different application scenarios. Built on this premise this chapter presents ideas for today's digitalized enterprises, which seem to disregard the fact that the latest technological advancements enable different ways of interacting with computerized systems, and that current interaction methods are bound to constraints of half a century ago. Given today's computational power, the engineers of former decades would not have had to invent intermediary interaction devices such as the mouse, if direct manipulation with touch screen or natural language processing had been possible. The possibilities for modern enterprises to overcome the restrictions of interaction devices from the past are considered.

Keywords Human-computer interaction · Digital enterprise · Digital innovation
Natural language processing · Optical character recognition · Spoken language processing

S. Jüngling (✉) · J. Lutz · S. Korkut · J. Jäger
Institute for Information Systems, University of Applied Sciences and Arts Northwestern
Switzerland, Peter Merian-Strasse 86 4002 Basel, Switzerland
e-mail: stephan.juengling@fhnw.ch

J. Lutz
e-mail: jonas.lutz@fhnw.ch

S. Korkut
e-mail: safak.korkut@fhnw.ch

J. Jäger
e-mail: janine.jaeger@fhnw.ch

© Springer International Publishing AG 2018 243
R. Dornberger (ed.), *Business Information Systems and Technology 4.0*,
Studies in Systems, Decision and Control 141,
https://doi.org/10.1007/978-3-319-74322-6_16

1   Introduction

Human-computer interaction (HCI) is a broadly researched field with a long history describing "the means of communication between a human user and a computer system, in particular referring to the use of input/output devices with supporting software" (Encyclopedia.com 2017). The main goal of these research areas is to achieve the most natural interaction of a person with a computer or machine.

When humans interact with computers the main senses used are vision, hearing and touch and the traditional interaction pattern is characterized by mechanical input through peripheral devices such as the mouse and keyboard. The feedback of the computer is visualized on a screen.

The evolution of HCI capabilities helps us to understand how we currently interact with computers. We focus on the importance of the context as well as the technical constraints in particular decades and reflect on human habits and interaction patterns in order to show the potential improvements of processes and the ways in which we will interact with computers in the digital enterprises of the future.

1.1   Historic View on Peripheral Interaction Technologies

In order to understand some of the constraints of current HCI, Fig. 1 provides a brief overview of the historical evolution of peripheral interaction technologies.

In the early days computers were developed to support a small number of technical specialists in storing and retrieving digitized data and managing limited sets of computational tasks within the scope of large enterprises. The interaction with mainframes was based on textual input via a keyboard and the output was displayed mainly as text on 3270 terminals without any application logic. In the 1960s, the first computers with electric keyboards were developed, and commands were immediately seen on a screen. Since the beginning of programming, a myriad of different programming languages have been developed to improve the way in which humans— *specialists with keyboards*—teach computers to do calculations and automate tasks. To efficiently manage structured data in relational database systems, the structural query language (SQL) was designed with the aim of formulating queries in human-friendly short SQL statements—at least a little human centric.

When computers became *personal computers* in the 1980s, computational power was used by a much broader population, mainly in the area of office tasks and back-office automation. Instead of terminals, windowing systems were established. Although the invention of the mouse itself happened around the same time as the keyboard, and was way ahead of its time, pointing devices were now needed to point to specific locations in documents or manipulate windows. This led to the breakthrough of the mouse as a consumer device.

In the middle of the 2000s, the computer industry turned towards the entertainment and music sector, raising the demand for *highly mobile devices*. Attracting a larger

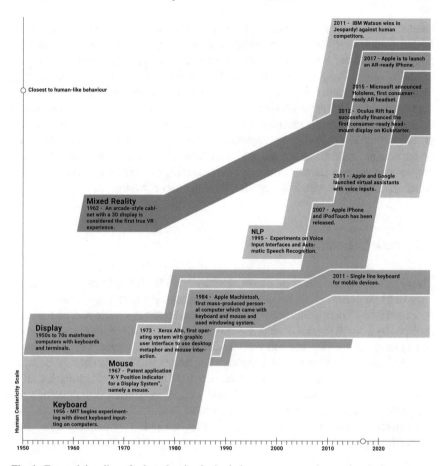

Fig. 1 Focused time line of selected technologies in human-computer interaction devices

audience with less technological knowledge thus required even simpler and handier input solutions. The industry's answer was the introduction of the touch screen and the mainstream adoption of iPods and smartphones. Tablets mainly use touch screens for input instead of keyboards. Touch screens have since then been a widespread means of human-computer interaction, representing input and output devices at the same time, and being used by natural pointing gestures with intuitive eye-hand coordination and direct visual feedback—which is much more *human centric*.

1.2 Historical Constraints

With current technologies humans are capable of using fingers, voice, eyes, head and body position to interact with computers and the circumstances of HCI have continuously changed over the last decades. Smart phones are used preferably in a single handed, thumb-based interaction style. The question arises as to why smartphones have the same interface for writing text as the first keyboards that were introduced. Although the layout is well-known, the keys are too close to each other for the fingers of many people. People are still learning how to use the keyboard more efficiently, instead of the computer learning more efficiently how to support humans in the writing task. It should be a primary target to *optimize the interface from the perspective of the human*, instead of the machine. There are still constraints, but they certainly are different from the original constraints such as the QWERTZ/Y keyboard layout based on mechanical constraints that the engineers faced at the time of their invention. We suspect that the constraints come from the technology of the tools, the context and situations in which the tools are used or from the *users' habits*.

1.3   New Solutions

The digital enterprise has huge potential with regard to HCI to transform unimodal peripheral devices such as the keyboard and mouse towards more human-centric user interfaces. Traditional external peripheral computer devices could be partially replaced by a conglomerate of sensors providing the ability to integrate touch, speech and gesture recognition (Gorecky et al. 2014) in more intuitive ways. The main conscious channel for human communication is language. We consider how far away are we from using *natural language processing* (NLP) as the basic pillar for HCI. This chapter examines current states of hand written and spoken language analysis, especially the process of converting these input data into processable pieces for a computer. We propose some future application scenarios for the language-based interaction approach.

2   Challenges of Human Language in HCI

For a machine to comprehend human language, many layers are needed, such as morphology, syntax, semantics and pragmatics. Morphology and syntax are concerned with the basic units of words known as morphemes, and how they form words and eventually correct sentences. Semantics addresses the creation of logic and real-world references in words, sentences and text. Pragmatics discusses the issue of how sentences are used in different situations and how to interpret them (adapted from (Allen 1995)).

Generic natural language processing (NLP) consists of a data input processing artifact, an application artifact containing the logic to be applied to the input data and an output artifact that makes the processed data interpretable for humans. Any of these artifacts are likely to draw some kind of knowledge from the afore mentioned subareas.

In NLP, a clear distinction is made between *written versus spoken* language. Written texts are well structured, in the sense of well-defined with complete sentences, coherent threading and an absence of dialects. However, spoken language or hand written notes first need to be converted to proper written language before universal NLP tools make them machinable. Human talking might end up unfinished, mispronunciations are quite common, dialects are sometimes hardly comprehensible, even for humans, and written letters might be illegible. This opens up completely new issues on a metalevel: small versus large vocabulary, clean versus noisy language, read versus spontaneous speech and single versus multiple languages (Yu and Deng 2014).

3 Handwriting and Optical Character Recognition

Text recognition has its roots in the field of optical character recognition (OCR). In the early days, special types of fonts such as OCR-A were invented to improve the recognition of machine text in pixel-based images and make the process less error prone. In the early days, optical mark recognition helped to process hand-written paper-based forms, which is one of the initial application domains. Intelligent character recognition further improves the recognition rate by using additional information about the text. For instance in order to recognize ZIP codes, it can verify the input by mapping it with existing ZIP codes. *Highly-automated address recognition* in mail distribution centers is possible for all sorts of different handwriting styles. The efficiency of modern logistic centers would not be possible without high-speed scanners that read and recognize hundreds of scans per minute. Different application domains for OCR tools are mobile phones, tablets and notebooks where handwriting and subsequent text recognition is used as an alternative input method instead of the keyboard. While batch processing in big scanning centers is quite established, the usage of handwriting and text recognition for end-users on mobile devices is still not very common and quite different.

3.1 Language Translation and Digital Learning Assistant

There is a wide range of different tools as well as architectures for handwriting and text recognition in the application domain of learning foreign languages, where mobile phones provide invaluable extensions to human communication skills. Figure 2 shows a typical situation, where the mobile phone serves as a practical, omnipresent translation tool. We can differentiate between offline and online scenarios. OCR software used on tablets or smartphones may connect to a back-end system, where more CPU power permits the use of sophisticated image and text recognition systems to process the input. On the other hand, offline recognition is very important

Fig. 2 Mobile translation support

to enable it to be used in many practical situations where Internet connectivity is not reliable or too expensive.

In such situations, all the processing needs to be done locally. Due to less CPU capacity, reduced sets of word collections and less complex but more efficient algorithms have to be used. However part of the performance gain resulting from powerful back-end systems is lost due to network latency time of the client server architecture. According to Keysers et al. (2017) Google's multi-language handwriting recognition system currently supports 22 scripts and 97 languages and can be used as both an online as well as offline translation system. The system was originally designed for online handwriting recognition, which takes about 40 ms per recognized English character including server round-trip time, which compares to 25 ms per character for on-device recognition on a LG Nexus 5 phone. Although the recognition rate is reported to be between 10–40% lower than the cloud recognizer, this offline performance is quite impressive. Such offline translation capabilities provide great help in many practical situations and are valuable extensions to human translation skills.

Independent from the different application scenarios, a considerable amount of research is invested in the different methods as well as different scripting alphabets such as Latin, Arabic, Cyrillic, Chinese or Indian, just to mention a few. Obviously enough, recognition rates are different for the interpretation of single characters and for entire handwritten texts. A recent overview of Arabic handwritten character recognition using Deep Belief Neural Networks by Elleuch and Tagououi (2015) reported an error classification rate of 2.1% for the standardized training single character database HACDB (Lawagli et al. 2013) and an error rate of more than 40% for the standardized word level database ADAB with training data. Although similar standard database sets exist for other languages such as Chinese and English, results are difficult to compare due to the different number of words or character sets. Wu et al. (2015) recently published Chinese character recognition rates for CASIA-HWDB and ICAR-2013 competition data sets of around 95% for different neural network based language and shape models.

Fig. 3 Character writing tutoring applications

Also in the context of foreign language support, the mobile device might be a valuable aid to teach handwriting in foreign languages with different character sets. *Small applications* such as Chineseskill (2017) provide good instructions in terms of presenting the correct stroke sequences of the Hanzi Characters, as shown in Fig. 3 where a person receives instructions how to write the Chinese characters with the strokes in correct order.

The learner may not only exercise the correct stroke sequences and receive feedback where the recognized written input differs from the standard order, but may also listen to the pronunciation of the words. There are exercises to say short sentences and the system gives feedback in terms of words written in green or red letters for correct or wrong pronunciation. Additional gamification elements such as embedded flashcards or quizzes help to keep the learners on track. In this way the mobile phone can be used as a valuable multi-functional I/O device for visuals and sound. Although the tip of the finger can be used for drawing for such small applications, pen-based input devices have the potential to replace the mouse as graphical interaction and pointing devices.

3.2 Electronic Notes and Signatures

There are two particular application scenarios where handwriting is still very close to traditional paper-based processing. The first of these is handwriting in the form of adding handwritten marks to existing documents, secondly human signatures in the context of authorization and processing of tasks. In both cases, there are two fundamentally different solution approaches for the equivalent paperless processes.

The traditional solution sticks to the paper paradigm, the second tries to find digital-native solutions. With the help of the touch screens of modern tablets and notebooks, electronic pen-based markups are as efficient as traditional pen-based markup on paper, and eliminates media breaks in many business processes. Electronic pens, in some cases magnetically attached to the side of the screen, can be used comfortably to make notes on documents. Traditional metaphors such as placing an eraser on the upper side of the pen or a double click on the back-end-button resulting in opening a new digital note on the computer screen might be used to transfer the traditional usage patterns into the digital scenario. It would even be possible to detect whether the swipe on the document on the screen comes from the human hand or the tip of the pen, resulting in either a page change event or in a thick red mark across the page. Some software tools already support this distinction while others do not, which still makes handling laborious in many practical situations. Another example of similar usability shortcomings is how to carry out an "undo" action. When typing on a keyboard, it is quite common to quickly press Ctrl-Z, but shortcuts for an undo action with a pen-input-device still remain to be established and standardized.

When we consider digital solutions, an electronic signature that authenticates a person's signature for authorization purposes not only eliminates the media break in processes, but also provides additional functionality in terms of proving the originality and authenticity of the signed document. Regarding the process of highlighting important information in digital documents, we might also consider solutions where part of the processing and visualization is carried out by the computer behind the scenes instead of by humans. Well-known examples are spell checkers, which highlight wrong spelling. However, some intelligent algorithms might be used to think up more elaborate visualizations, which do the highlighting dynamically based on certain keywords of interest or which show the keywords in the context of some knowledge graphs, which can then be explored in more intuitive and more interactive ways.

Finding innovative HCI solutions for a particular situation not only depends on the person's preference but also on the context and location in which humans work. For many people, handwriting is slower than typing on a regular-sized keyboard. Nevertheless, in some situations, handwriting can be more convenient than keyboard-based text input. It is a different situation if notes are taken by a mechanic during a maintenance activity, by a student during math lectures at universities or by a manager during a meeting. Handwritten notes on a touch screen can be transferred on the on the spot into machine-readable characters and are, therefore, researchable assets providing tremendous advantages over traditional paper-based notes. In increasingly rare situations, paper-based notes which can be scanned later on might be the optimal solution.

3.3 Written Pattern Recognition

Aside from education and language support applications, other services could be developed that are based on image processing and pattern recognition using mobile phones. Pictures could be selected and sorted, based on facial recognition of persons. Cars could be identified based on the automatic detection of license plates, which was demonstrated in Malaysia in 2012 (Mutholib et al. 2012) and applied in different situations from speed prosecution to the management of parking lots.

4 Spoken Language Processing

Spoken language processing (SLP) is a specialized research area of the field of NLP. It is considered a bridge to improve human-human communication (HHC) and human-computer communication (HCC) (Yu and Deng 2014). In the late 1970s a survey of the available systems found another reason for failure: interfering noise (Lea 1980). Although the term "robust speech recognition" emerged, the original shortcomings remained. However, the post-millennial technological improvements sparked interest in the topic of NLP once again and the first effective implementations of unsupervised learning algorithms were demonstrated. More concretely, SLP engages with the improvement of three generic application scenarios (Huang et al. 2001): Spoken language interface and knowledge partners, which are both applied in HCC, and speech-to-speech translation, which represents the HHC aspect.

- The *spoken language interface* encompasses the idea of replacing current interface concepts such as the mouse, keyboard or touchscreen. A simple command like "disable wireless" or "increase letter size" is much more natural than clicking through several menus to perform the simple, one-dimensional task of updating a machine setting.
- *Knowledge partners* are much more complex. The most advanced knowledge partner today is IBM's Watson, which defeated two human champions in the quiz show Jeopardy (Gabbatt 2011). These artificial systems are able to accomplish any of todays human tasks and thus rely on extensive knowledge about the world combined with sophisticated search, reasoning and communication capabilities. Todays more affordable solutions are Intelligent Personal Assistants (IPA) which assist human beings in their every day life. There are two basic archetypes of IPAs: with internal and external initiation. Today's assistants (e.g. Siri or Cortana) are initiated externally, by either pressing a button or activating them via voice commands, while agents with internal initiation permanently gather data about a customer and his or her environment to interfere if an obstacle arises (e.g. a traffic jam en route).
- *Speech to speech translation* is a bi-directional communication system. After processing and understanding the input sentences, the machine translates them into a different language and outputs the content in spoken language. This domain made revolutionary progress lately in the private sector, where proprietary software and devices are capable of translating languages back and forth instantly.

A spoken language system (SLS) implements one of the above mentioned application scenarios. According to the literature, these types of systems consist of at least one of the following three functionalities (Huang and Deng 2010). Some separate the application logic from the spoken language understanding system and propose to form their own subcomponent, known as a dialog manager (Yu and Deng 2014). For the sake of simplicity, the densest version is considered:

- *Automated speech recognition*, which converts audio signals to words.
- *Spoken language understanding*, which can give the previously converted words a meaning. Furthermore, it is responsible for the logic of handling possible queries contained in the input and executing tasks that are required to answer the queries.
- *Text-to-speech*, which is capable of converting text back to speech.

Figure 4 shows a generic architecture of an *automated speech recognition component*. It converts sound waves into feature vectors, forwards this to the decoder to search through acoustic and language sources to find the most likely word sequence match to a given input vector (Karpagavalli and Chandra 2016). The acoustic front end creates a feature vector as a compact representation of the input signal (Anusuya and Katti 2011). Many different feature sets and methods are available, but the most common is the mel frequency cepstral coefficient (MFCC). These predefined feature sets (known as feature domains) have lower computational costs compared to the model domains. Model domains incorporate noise into their acoustic models and eventually achieve a higher degree of accuracy, but at a significantly greater computational cost (Li et al. 2014).

The resulting vector is then passed on to the search algorithm, which runs matching and verification techniques against its three main sources:

- The *acoustic model* represents acoustic features for phonetics which can be queried by the decoder. It usually consists of some sort of a Hidden Markov Model (Baum 1972; Huang and Deng 2010; Li et al. 2016).
- The *language model* represents common language rules. It allows the system to verify that matched phonetics, words and sentences are correct in a general sense. In most cases it is either a formal specification or a stochastic language model (Jurafsky and Martin 2000).
- The *lexicon* contains all the possible words in a given language.

The search strategy mostly employed for speech recognition in the decoder is the Viterbi Algorithm (Viterbi 1967; Huang and Deng 2010; Karpagavalli and Chan-

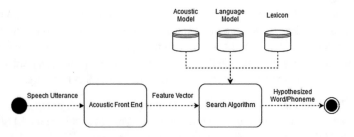

Fig. 4 Speech recognition component architecture (adapted from Karpagavalli and Chandra 2016)

dra 2016). However, due to challenges with larger vocabularies, it was extended with beam search, extended Viterbi and forward-backward algorithms (Jurafsky and Martin 2000).

The *spoken language understanding component* is the most debatable and most adaptable of the three components. It is supposed to understand words and sentences received from the automated speech recognition component, derive semantics and put it into a dialog context to generate a set of tasks that satisfies the purpose of the conversation and maintain some kind of discourse strategy (Huang et al. 2001; Yu and Deng 2014).

The *text-to-speech component* converts written text back into meaningful and understandable speech. The incoming text is normalized and undergoes a phonetic analysis to convert it into the corresponding phonetics. These are enriched by a prosodic analysis that adds pitch, duration and further pronunciation information to the sequence. Eventually, a speech synthesizer component generates the sound waves (Huang et al. 2001). Nowadays, there are several proprietary products available, such as Google Translate and Skype Translator.

4.1 Vocal Navigation in Simulated Realities

Virtual Reality (VR) and Augmented Reality (AR) are two state-of-the-art computerized simulation media where the reality is simulated up to an indistinguishable and hypothetical trueness. In VR simulations, digital elements are placed in a simulated world and viewers experience the simulation in full immersion with the help of a head-mounted display, whereas, in AR simulations, digital elements are placed as a layer on top of the real-world environment through a holographic display glasses or smart phones. In AR, the viewers are confronted with a sensory duality between the physical and artificial phenomena that are perceived in the same three-dimensional space.

AR/VR applications are becoming more popular in various business sectors such as finance, surgical training, or for robotic navigation, education or in maintenance dashboards to provide a visual representation of maintenance tasks as investigated in Probst (2017) for the usage of AR in the service sector of ABB Turbo Systems. As shown in Fig. 5, real-world machinery can be augmented with additional technical details such as important tolerance metrics, which need to be guaranteed while following detailed mounting instructions.

Other application scenarios are found in architecture/real estate sectors, where they project 360° experiences of a house, a village or landmarks on VR and users can walk through virtually. These applications are also on the radar of artists who showcase their work so that art enthusiasts can walk through a virtual gallery (Halsey 2016). VR is also applied in tourism, as it enables users to take a walk in a completely new city or country, or look at heritage sites from the comfort of their homes (Safrudin et al. 2015).

Fig. 5 AR-animated tolerance metric instructions

In a recent research project (Brennenstuhl 2017) explored the possibility of walking through virtual reality applications on a smartphone by using voice commands. The usage of the smartphones instead of sophisticated VR technology was a market penetration. Current VR setups that allow for interaction with wireless controllers are not widespread and mostly used by gaming enthusiasts. Brennenstuhl proposes that users in virtual environments should be able to freely choose where to go and at what speed with hands-free devices based on travel and path-finding. *Travel* describes how users move something or themselves through the world (whether real or virtual), while *path-finding* studies describe how users find out where they are and where they need to go (Craig et al. 2009). The tests and corresponding questionnaire showed that steering by voice commands can enhance navigation in VR. However, the current limitations of speech recognition methods (lack of stability, background noise and the need to press a button for activation) still need to be solved.

4.2 Audio Message Recording Replaces Typed Chat

An interesting trend is occurring in the technologically-driven field of private communication. WhatsApp users have the alternative to either use voice chat or transcribe voice messages into text before sending. In the Asia Pacific region, where WeChat is more commonly used than WhatsApp, it is quite common that the entire communication is done by voice messages. This may be due to the fact that text-based chats are based on Pinyin, a "phonetical" representation of their traditional Chinese Hanzi characters.

5 Conclusion

Given the current status of technology, our HCI patterns still largely rely on constraints from the past and have not receive the major overhaul they deserve. Operating systems engineers started to increase accessibility and added voice control mainly for visually impaired humans. However, with the establishment of tablets and smartphones, a first step towards a new way of interacting with computer systems was introduced. Pointing with one's hand instead of a mouse is much more intuitive, but talking instead of using a keyboard is still in its infancy. Many tools for language analysis, spoken or written, still lack precision. With the penetration of virtual and augmented worlds with head-mounted devices it is even more obvious that we have new constraints and visual restrictions and can no longer rely on the mouse and keyboards. Augmented reality has also started to implement gesture controls. Gesture control is a great achievement for audio-impaired people but the main communication channel of humans is the voice. Visualizations by virtual and augmented realities will find their way into corporate offices, production facilities and enhance client experiences in the real estate market or in the purchasing of goods. Language controls would not only be most beneficial in such applications. It is also time to reposition the role of voice-based control in the digital enterprise and change the workplace constraints accordingly.

References

Allen J (1995) Natural language understanding, 2nd edn. Benjamin-Cummings Publishing Co., Inc., Redwood City, CA, USA

Anusuya MA, Katti SK (2011) Front end analysis of speech recognition: a review. Int J Speech Technol 14(2):99–145

Baum LE (1972) An inequality and associated maximization technique in statistical estimation for probabilistic functions of markov processes. In: Inequalities III: proceedings of the 3rd symposium on inequalities. Academic Press, New York, pp 1–8

Brennenstuhl M (2017) Using voice commands to enhance the navigation in virtual reality on a smart phone. Unpublished, master thesis

Chineseskill (2017) Chineseskill–learn Chinese language [mobile application software]. http://www.chinese-skill.com/cs.html. Accessed 27 June 2017

Craig AB, Sherman WR, Will JD (2009) Developing virtual reality applications: foundations of effective design. Morgan Kaufmann

Elleuch M, Tagououi N (2015) Arabic handwritten characters recognition using deep belief neural networks. In: 12th international multi-conference on systems, signals and devices

Encyclopediacom (2017) Human computer interface. In: A dictionary of computing. http://www.encyclopedia.com/computing/dictionaries-thesauruses-pictures-and-press-releases/human-computer-interface. Accessed 29 June 2017

Gabbatt A (2011) IBM computer watson wins jeopardy clash. https://www.theguardian.com/technology/2011/feb/17/ibm-computer-watson-wins-jeopardy. Accessed on 06 June 2017

Gorecky D, Schmitt M, Loskyll M (2014) Mensch-Maschine-Interaktion im Industrie 4.0-Zeitalter. Springer Fachmedien Wiesbaden, Wiesbaden, pp 525–542. https://doi.org/10.1007/978-3-658-04682-8_26

Halsey E (2016) 5 ways virtual reality will change architecture. http://archsmarter.com/virtual-reality-architecture. Accessed on 29 June 2017

Huang X, Deng L (2010) An overview of modern speech recognition. In: Damerau F, Indurkhya N (eds) Handbook of natural language processing. Chapman Hall/CRC, pp 339–366

Huang X, Acero A, Hon HW (2001) Spoken language processing: a guide to theory, algorithm, and system development, 1st edn. Prentice Hall PTR, Upper Saddle River, NJ, USA

Jurafsky D, Martin JH (2000) Speech and language processing: an introduction to natural language processing, computational linguistics, and speech recognition, 1st edn. Prentice Hall PTR, Upper Saddle River, NJ, USA

Karpagavalli S, Chandra E (2016) A review on automatic speech recognition architecture and approaches. Int J Sig Process Image Process Pattern Recognit 9(4):393–404

Keysers D, Deselaers T, Rowley HA, Wang LL, Carbune V (2017) Multi-language online handwriting recognition. IEEE Trans Pattern Anal Mach Intell 39(6):1180–1194. https://doi.org/10.1109/TPAMI.2016.2572693

Lawagli A, Angelova M, Bouridane A (2013) HACDB handwritten Arabic characters database for automatic character recognition. In: Proceedings of the 26th annual ACM symposium on user interface software and technology, pp 531–538. http://ieeexplore.ieee.org/stamp/stamp.jsp?arnumber=6623974

Lea WA (1980) The value of speech recognition systems. Trends in speech recognition. Prentice Hall, Upper Saddle River, pp 3–18

Li J, Deng L, Gong Y, Haeb-Umbach R (2014) An overview of noise-robust automatic speech recognition. IEEE/ACM Trans Audio Speech Lang Proc 22(4):745–777

Li J, Deng L, Haeb-Umbach R, Gong Y (eds) (2016) Robust automatic speech recognition. Academic Press, Oxford

Mutholib A, Surya TG, Kartiwi M (2012) Design and implementation of automatic number plate recognition on android platform. In: International conference on computer and communication engineering

Probst T (2017) Einsatz von augmented reality im servicegeschäft von ABB turbo systems. Unpublished, bachelor thesis

Safrudin N, Fay M, Changa A, De Wit B (2015) Game changing digital technologies. https://www.researchgate.net/publication/308694636_Game_Changing_Digital_Technologies. Accessed on 29 June 2017

Viterbi A (1967) Error bounds for convolutional codes and an asymptotically optimum decoding algorithm. IEEE Trans Inf Theory 13(2):260–269

Wu YC, Yin F, Cheng-Lin L (2015) Improving handwritten Chinese text recognition using neural network language models and convolutional neural network shape models. https://doi.org/10.1016/j.patcog.2016.12.026

Yu D, Deng L (2014) Automatic speech recognition: a deep learning approach. Springer Publishing Company, Incorporated

Prototype-Based Research on Immersive Virtual Reality and on Self-Replicating Robots

Rolf Dornberger, Safak Korkut, Jonas Lutz, Janina Berga
and Janine Jäger

Abstract This chapter presents our recent research in the field of virtual reality (VR) and self-replicating robots. The unifying approach lies in the research philosophy of using consumer market gadgets, mostly developed for the gaming and entertainment business, in order to design and implement research prototypes. With the prototypes, our research aims to better understand real-world problems and derive practice-oriented solutions for them. In the field of VR, these prototypes are dedicated to identifying new business-relevant use cases in order to provide an additional benefit for business and society. A wide range of examples, such as claustrophobia treatment, financial data analysis, gesture control and voice navigation are discussed. In the field of robotics, the idea of self-replicating robots governs particular research questions. Here, the focus is on using model prototypes enriched with artificial intelligence for indoor navigation, computer vision and machine learning. Finally, the prototype-based research approach using gadgets to produce results is discussed.

Keywords Prototype-based research · Gadgets · Virtual reality
Self-replication · Robots · Artificial intelligence

R. Dornberger (✉) · S. Korkut · J. Lutz · J. Berga · J. Jäger
Institute for Information Systems, University of Applied
Sciences and Arts Northwestern Switzerland, Peter Merian-Strasse 86,
4002 Basel, Switzerland
e-mail: rolf.dornberger@fhnw.ch

S. Korkut
e-mail: safak.korkut@fhnw.ch

J. Lutz
e-mail: jonas.lutz@fhnw.ch

J. Berga
e-mail: janina.berga@fhnw.ch

J. Jäger
e-mail: janine.jaeger@fhnw.ch

© Springer International Publishing AG 2018　　　　　　　　　　257
R. Dornberger (ed.), *Business Information Systems and Technology 4.0,*
Studies in Systems, Decision and Control 141,
https://doi.org/10.1007/978-3-319-74322-6_17

1   Introduction

As stated by Hotho and McGregor (2013, p. 1), the games industry is young, evolving and "frequently experiencing disruptive innovation". Society is in the midst of technological transformation, and the game industry is its driver: At an amazing speed, established companies and start-ups are developing and producing new technologies in the form of gadgets that can either burst like a bubble (as some do) or penetrate our lives in a way that no one would ever have expected and become normality. An older example—and older means 10–15 years here—is the transition from basic cellphones to powerful smartphones; in a very short time, smartphones have become a solid part of our social lives. Society has come to a turning point where it is almost impossible to imagine not having a smartphone at hand—as a connector to the Internet capable of instantaneously finding and processing different types of information. Bearing this in mind, it is not unnatural to presume that the same thing might happen to other gadgets such as smartwatches, smart glasses or autonomously driving cars. (The word *smart* seems to be a good indicator for that.)

According to the Merriam-Webster Dictionary (Merriam-Webster n.d.), a *gadget* is "a small electronic device with a practical use". Whilst new technologies are being introduced to consumers at a very fast pace, gadgets are becoming increasingly smarter as a result of continually being online thanks to the *Internet of Things (IoT)* and different network protocols (i.e. Bluetooth, RFID[1] tags, and so forth) that support interconnectivity between gadgets within the ecosystems. Gadgets address needs that are not necessarily a technological challenge, and often support users by means of features such as automation, connectivity, immediacy, time-saving and hassle-release aspects, to name but a few. There is great potential for smarter gadgets and innovative industry applications for an automated future scenario where gadgets are designed to fulfill our unexplored needs. Gadgets build a bridge between the user and the provider, and just like any other digitalized product or service, the experiences themselves also provide valuable data about the user. Here, gadgets play the role of smart agents between systems and their users. This means that all input can be collected and user behavior can be tracked and analyzed in order to assist, optimize or mediate the use of the service.

In this book chapter, we describe our research strategy, where we advance from merely researching the usage of a new technology to its potential business applications. With the fast-paced advancements in the development of IT, communication technologies and new technologies, our applied research aims to develop business cases to implement the use of new technologies in real business scenarios (e.g. as IoT in Industry 4.0) rather than to research the gadget technology by itself in depth. Thus, we follow the strategy of designing and building prototypes and designing experiments to investigate the performance of selected technologies (e.g. virtual reality using particular consumer (gaming) products, or robotics using LEGO Mindstorms EV3 robots) in other settings than the gaming and entertainment sectors. Therefore, we build models of real-world scenarios or problems on the scale of laboratory

---

[1]Radio frequency identification.

experiments. Using particular gadgets, we discuss the problems modeled and implement prototypes that produce results, which we then try to transfer back from the model level to the level of real-world problems. Thus, we address research questions by developing prototypes and discussing the generated solutions.

Moreover, to support these prototypes and give them a chance to develop as sustainable businesses, we apply asymmetric business models, which are business models that cross industries and amalgamate existing industries with new opportunities and innovation (Vakulenko 2013). In other words, we seek to discover new visionary products or services and to optimize existing ones by redefining the original usage of gadgets (mostly from the area of gaming, entertainment or lifestyle) to new cases. Technological development is often tricky to predict. Gartner's Hype Cycle (Gartner n.d.) is a good example of this. Gartner's methodological approach represents the maturity and adoption of emerging technologies in a visual graph, ranging from *Innovation Trigger, Peak of Inflated Expectations, Trough of Disillusionment, Slope of Enlightment* to the *Plateau of Productivity* (see Fig. 1). Gartner releases a yearly prediction of new technologies and shows how they evolve over time; however, new business opportunities are usually not noted. As if out of nowhere, asymmetric business models can create powerful companies in the market. Some examples of early adoption in the industry applications are already available, e.g. for the use of virtual reality: (1) retail—presenting products to customers, virtual shopping (Lui et al. 2007), (2) health—phobia treatment (Carlin et al. 1997), and (3) education—immersive learning environments (Dede 2009). Tech giants, such as Google (2017) and Facebook (2017), are now producing experiences with virtual reality and in autonomous navigation. More and more start-up companies are reaching newer audiences and markets as they explore new sectors and services through technology-oriented interactions and new gadgets.

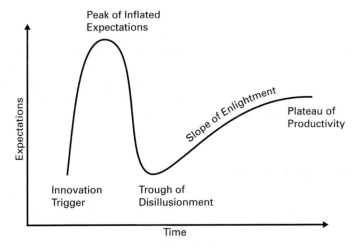

Fig. 1 Gartner's Hype Cycle, adapted from (Gartner n.d.)

Nonetheless, technology readiness, or as mentioned before, technology acceptance, has yet to be established for leading enterprises that are investing in creating digital platforms. Gartner's Hype Cycle is a well-established approach that is used for investment and risk management (Gartner n.d.). Companies can use this approach to guide them in decision-making but there is no guidance available for the actual implementation strategy and identification of use cases within Gartner's Hype Cycle method: Each phase of the Hype Cycle refers to the life of a specific technology but it is not clear how to integrate/implement the technology in an existing business model. For example, in the 2017 release of Gartner's Hype Cycle (Gartner 2017), virtual reality appears on the *Slope of Enlightment* (see Fig. 1). In this phase of the Hype Cycle, enterprises often launch pilot products and invest in the development of use cases.

Similarly, in academic research, increasingly sophisticated studies are being published, analyzing developed use cases. For instance, applying virtual reality in a shopping experience, the walking route taken by shoppers in a supermarket (including every move they make, even just their head movements) can later be reconstructed and data studied to optimize the services (Altarteer et al. 2016) as well as for user analysis and customer experience improvement (Bigné et al. 2016). Such studies enable enterprises to envision the potential and opportunities for their business cases. Once technology acceptance has been achieved, other use cases will evolve more easily. Building on the previous example, the connection between the user data and the service is crucial to harvesting the underlying knowledge for the analysis of human behavior. Data collection from these gadgets in combination with machine learning algorithms can provide new ways of understanding information. In this context, users continuously provide data to the system. This data is collected, analyzed, compared with larger data sets and certain predictions or findings are contextualized as information. Finally, the information is returned to the user as knowledge (Hey 2004).

In the following section, we focus on virtual reality (VR) with particular attention to our prototype-based research to identify business-relevant use cases. Firstly, we present results from our research on VR applications and develop business scenarios and new ways of interacting. Secondly, we discuss our findings that confirm our vision of developing business cases supported by smart gadgets. We believe that this new genre of applications has great significance for industrial applications and their true potential is still to be discovered.

2 Virtual Reality Use Cases

Virtual reality is an immersive medium which is often referred to as computer-simulated reality. In VR, the medium replicates a digital environment, allowing the user to interact within the endless boundaries of digital cyberspaces. In contrast, augmented reality (AR) applications integrate digital imagery as supplementary input on the physical real-world environment, as, for example, the laws of physics still apply to the user interaction. In other words, VR and AR refer to the technologies that merge reality and virtualization and connect these two spaces in a simulated or real environment (Milgram and Kishino 1994). When we discuss these particular mediums, our main reference point is "head-mounted displays" (HMDs), i.e. the devices that provide information on demand in an AR or VR environment. As early as 1998, Watts et al. (1998) presented a vision for companies showing how this virtual reality technology can be used to be more innovative. Head-mounted displays (in VR or AR) possess great potential for business through the innovative use cases and opportunities in different industries as well as the new forms of interaction offered.

VR applications are becoming increasingly popular in various business sectors, e.g. tourism, real estate, retail, education, health care, robotics, finance and maintenance (Huang et al. 2016; Barnes 2016; Stansfield et al. 2016; Weise and Mshar 2016). For example, just by wearing an HMD, one can take a walk in a completely new city or country, or look at heritage sites in the comfort of one's home (Safrudin et al. 2015). In addition to VR, there is also 360° video technology, which allows to create a fixed-view angle in only one direction to experience a spherical projection as a "look-around" view. Applications with a virtual 360° walk-through possibility are also on the radar of artists who showcase their work in a way that art enthusiasts can walk through a virtual gallery (Halsey 2016).

2.1 Early Business Cases—From Fiction to Reality

In a recent study, we analyzed popular science fiction movies to find some HMD applications and selected from them some hypothetical and technologically achievable business use cases (Doganci 2015). We divided the use cases using AR wearables into three categories: (1) edutainment, (2) sports, e.g. navigation and access to vital information, and (3) emergency services, e.g. providing on-demand live information to rescue teams or fire fighters. In VR wearables, the use case examples were selected as (1) 360° games and videos in the entertainment category, (2) human gait analysis for the improvement of physical performance in the sports category, and (3) virtual emergency simulations for training purposes in the emergency services category. As a conclusion of this study, the current view of HMDs as an entertainment gadget does not hinder their high potential for effective business use cases. However, state-of-the-art technologies were found to be weak with regard to the display size, the complexity of the setup and possibly inconvenient user input. Nevertheless, with technological development and accessibility improvements, HMDs can achieve a better utilization.

One of the proposed use cases includes an AR prototype with a real-time information layer for business and educational tasks, which supports data-oriented knowledge transfer and information management. Furthermore, this proposal advances on information visualization and real-time analytics in complex business situations. In such systems, relevant information always remains in focus and insights are shared simultaneously within the visor of the AR display system.

In 2015, we started experimenting with the Oculus Rift development kit (OculusRift 2015) to see what was possible during the early stage of development. We examined and evaluated potential business cases: (1) Where VR would belong to the end customer and he or she would be able to use it from home (with all users receiving the same VR offering), and (2) where the Oculus Rift would belong to a company, would comprise customer-specific data (i.e. product demos) and would be ready for the customer (Lavanchy 2015). At that point, there were very limited smartphone-ready VR solutions and so, in order to ensure an earlier return on investment, we suggested development of more personalized VR products that would support the growth of a company product portfolio. It was also clear that the upcoming "VR-wave" would initiate opportunities for start-ups and small and medium enterprises to become specialists in this area (Lavanchy 2015). In order to analyze the use cases, we developed a Cartesian canvas (see Fig. 2): (X-axis) Passive user (i.e. only head movements on a predefined course) to Active user (i.e. free movement and interaction with controllers); (Y-axis) Real environment (i.e. 360° videos and panoramic photographs) to computer-generated environments (i.e. 3D models). The idea was to apply this canvas as a vision board, thus, it was shared with our students and researchers to envision potential application fields for use cases.

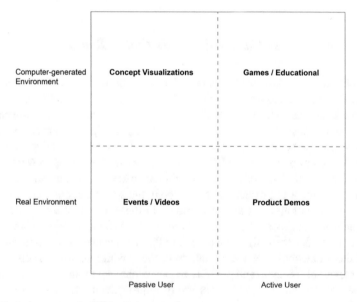

Fig. 2 Cartesian canvas for VR business scenarios

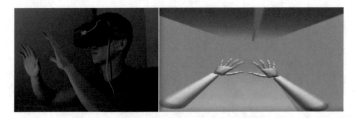

Fig. 3   Virtual reality as a treatment for claustrophobia (Buff 2015)

2.2   Virtual Reality Prototype for Treatment for Claustrophobia

In 2015, we developed a prototype to demonstrate a digital scenario that can be applied to treat claustrophobia (Buff 2015). It is possible to simulate a real-life scenario in virtual reality so that the patient does not need to imagine or face it in reality, such as spider phobia treatment experiment by Carlin et al. (1997). In effect, the presented method alone may become an obstacle for successful therapy because the sense of presence in the VR environment is not sufficient to explore the emotional impact on patients. Therefore, a further level of immersion is provided by including a gesture recognition device (LeapMotion 2010) attached on the HMD to render an immersive VR experience. This allows the users to see a virtual version of their hands and in this way, they can not only see the virtual surroundings but also interact with them as shown in Fig. 3.

In the context of this prototype, the ability to interact with the showcase was the key component. During the research experiment, the user would find himself in a long and narrow tunnel, and it would be possible to move the front wall of the tunnel closer to the user by just waving the wall towards himself with his hands, creating an even smaller space. In the moment of the highest level of discomfort, the user would be able to push the front wall away with his hands and open up the space. The prototype claims that it is possible to activate the users' own limits at their own pace, allowing them to practice being comfortable within confined spaces and feel more in control (see Fig. 3).

This research showed that VR can indeed be effective for therapy and that the used VR headset acts as an enabler in this scenario due to the considerably lower cost. Naturally, this prototype needs to be further developed and submitted to clinical testing and one must follow the technical advancements of the Oculus Rift or other VR systems, as each new release promises a greater number of possibilities. This scenario also illustrated how a user provides data to the system and, ideally, the data could also be studied to assess the effects of claustrophobia.

Fig. 4 Virtual reality for financial data analysis (Schär 2017)

2.3 Virtual Reality Prototype for Financial Data Analysis

Financial data has become a part of our everyday lives and we constantly seek more efficient and effective ways to store and analyze it. Gaining insight into our financial data in the form of categories is seen as a great step towards better financial planning and decision-making. To further this end, we explored a private e-banking scenario where we examined the possibilities of interaction with data in VR in a multi-dimensional way to see which 2D approaches to visualizing data were brought into the 3D in VR (Schär 2017). From the defined input methods (hand gestures, gesture controllers, speech recognition, physical placement of objects, full body tracking and 360° motion tracking), we utilized 360° motion tracking to enable the user's ability to move within the VR environment and interact using gesture controllers (see Fig. 4).

Essentially, through incorporating multiple linked views, we were able to confirm that it was easier to comprehend and interact with the data in virtual environments; however, the more complex queries were not yet easy to execute with such a prototype. As stated by Stone et al. (1994), "[a] visualization goal is to simplify the analysis of large-quantity, numerical data by rendering the data as an image that can be intuitively manipulated". Stone et al. define three important characteristics of VR in this regard: (1) VR exhibits high interactivity (the user's actions and the reactions caused are closely connected), (2) VR supports embodiment (the user is represented in the same spatial framework as the data), and (3) The VR representation is spatial in nature (all virtual objects are placed in a spatial framework). With this perspective, we foresee high research potential in exploratory data analysis via VR applications becoming part of a banking experience.

2.4 Prototype for Voice Navigation in VR Applications

Today, there is a vast amount of mobile VR applications. Those that permit the exploration of a place in VR can and are already being used, e.g. by real estate companies or by tourism agencies aiming at increasing sales by bringing their customers into VR prior to sale. In our recent project (Brennenstuhl 2017), we explored voice-commanded navigation for path-finding in VR environments, which, according to our results, can indeed enhance navigation in VR. However, due to lack of stability, background noise and the device requiring a button press for activation, there were certain drawbacks in the implementation of the technology in a wide range of applications at the time. Currently, two common methods are applied to navigate in VR environments: (1) Gaze-directed steering, where the user must look at a specific point for several seconds to select it and (2) target-directed steering, where the user is able to move within the VR environment into the direction in which she or he is looking. In our research, we developed a prototype application, which enabled both navigation types as well as voice-commanded navigation. We tested the prototype with participants who navigated within the application using all three options, and the voice-commanded navigation proved to be the most efficient in performing the required tasks. In fact, the majority of participants expressed their preference for using this method, should it be made more user-friendly. The study showed that there is clear potential in adding the voice command function to the VR navigation option for VR applications.

2.5 Prototype for Gesture Control in VR Data Visualizations

Big data is a term that has become part of the digital age. The success of many companies depends on their ability to process and analyze data, and, in many cases, this means vast amounts of it. Above all, VR provides new possibilities for data visualization and analytics, however, the question regarding the ways and extent of interacting with the data in VR remains.

Currently, physically interacting with VR is allotted to a game controller, or a remote with just a few buttons. By using the Leap Motion gesture control device combined with an HMD, one can also experience a touchless gesture-based interaction with data. This means that both hands are free of controllers and the user experience becomes more immersive. We conducted some experiments to explore how intuitively people can interact with information visualization in VR (Frey and Jurkschat 2017). Students and scientific staff took part in an experiment where we showed them a VR data visualization application that supported gesture control, and instructed them to intuitively come up with a gesture that would perform one of four given tasks (detail, on-demand, zoom-in, zoom-out and the overview). Interestingly, not all participants were able to come up with the right gestures to perform one task or another, while some participants showed the expectation of a functionality more similar to the use of a smartphone or a desktop computer. This encourages us to think that the defined gestures for VR can be learned and adapted with more frequent use of VR, and thus become intuitive.

As a result, data visualization in VR combined with touchless hand gesture control could provide a more immersive experience. In addition, some gestures were used more intuitively than others. Nevertheless, to develop a sustainable gesture standard, we recommended not basing it solely on the user's intuition but also considering the ergonomic consequences and possible technical limitations.

2.6 State-of-the-Art Developments of VR Prototypes and Use Cases

At the end of 2015, we forecasted that VR development would lead to a new social networking experience (Doganci 2015). About a year later, in spring 2017, Facebook launched its beta version of a social VR platform called Facebook Spaces (Facebook 2017). It is now possible to meet one's social network partners virtually in a virtual environment and explore various media (photos, videos) together. It seems to be possible to simulate a creative (i.e. business) environment on this platform by applying drawing tools. One can also join a virtual session in Facebook Spaces by video-calling a friend who is wearing an HMD and is in this environment—the caller then sees the other person's virtual appearance as an avatar and his or her virtual surroundings.

In the last few years alone, the capabilities of VR have developed exponentially, e.g. entertainment applications and speculations about the use of VR in industries such as health care, finance, education, etc. Throughout our research from 2015 to today, we have predicted that the real VR products that will affect the way we conduct business have still to be developed (Doganci 2015). The new consumer-ready VR HMDs are being introduced to the consumer's market. Indeed, the developer community has probably never been more active in developing new and innovative VR applications for a variety of industries. In order to achieve productivity, VR prototypes have to be developed and a variety of cross-industry use cases have to be identified both in industry- and academy-driven research projects.

3   Self-Replicating Robots

In this section, we present our vision for self-replicating intelligent robots, where we investigate the feasibility of related tasks by a simple robot kit. We share outcomes following our prototypes and their challenges to discuss serious business cases.

Computational power has increased tremendously in the last decade and this has led to a general resurgence of interest in the research domain of artificial intelligence (AI). AI systems can compete with human beings in challenging test settings. In 2011, IBM's Watson[2] defeated the reigning human champions in the television show Jeopardy (Gabbatt 2011). In 2016, Google's AlphaGo computer program beat the world's Go[3] champion (Russell 2017). Furthermore, the first intelligent personal assistants were released in 2011 with Apple's Siri (2011), Google's Assistant (2011), Microsoft's Cortana (2014) and Amazon's Alexa (2014), which have penetrated a small but growing portion of the consumer market. Further, Google have pushed their research in the area of self-driving cars, which have driven more than three million kilometers autonomously with few incidents (Bhuiyan 2017).

---

[2] A cognitive system enabling a new partnership between people and computers (IBM Watson 2017).

[3] A 3000-year-old Chinese strategy board game with a complex game play.

Within these perspectives, AI is finding more and more application potential in certain tasks where human input is no longer relevant. As they dehumanize these tasks, robots are facing another potential developmental challenge: self-replication. How can robots build themselves from scratch? Robots are being used to assist or replace humans in dangerous and repetitive tasks such as space exploration. Over the last few years, researchers have studied the feasibility of self-replicating robots (Mino 2015; Moses et al. 2014; Napp et al. 2006; Suthakorn et al. 2003). A common vision is that robots in autonomous robotic factories will be able to repair and replicate themselves. Such factories would require no control by humans from outside, although their state could be monitored. As mentioned, outer space exploration is another potential application (Chirikjian et al. 2002). Such self-replicating robotic factories placed on a planet could take advantage of the planet's resources to produce refined materials and energy, then provide the possibility to further explore and colonize space, and collect solar energy for terrestrial applications. Due to unresolved technical difficulties, however, these systems are not yet feasible. In the next sections, we will describe our research, which aims to take a step closer towards robotic self-replication. In each iteration, we focus on a particular aspect, which will increase the complexity by one level.

3.1 Prototype: Computer Vision Principles Towards Self-Replicating Robots Use Case

With the increase of computational power, there has been a resurgence of previously unsuccessful areas (Wolfram 2015; Fei-Fei et al. 2006; Fei-Fei 2015; Wang et al. 2015), which has successfully challenged the status quo. However, most computer vision (CV) applications have a very specific task and are far from an all-purpose CV system. In a particular research project dealing with the self-replication phenomenon, we examined computer vision principles using only low-cost hardware (Mino 2015). Using a Lego Mindstorms EV3 set, we designed and developed a modular robot prototype (see Fig. 5). It was thought that the robot would be able to replicate itself by assembling its modules. It can move in an unstructured unknown environment, where resource modules are spread randomly. Using a gripper to gather the modules and equipped with a camera, the robot operates autonomously through a remote software application. Then, the OpenCV library[4] is used to acquire, analyze, process and recognize objects in the video frames. A color-based object recognition algorithm was used because the resource modules were differently colored LEGO bricks.

In principle, a complex modular robot can hypothetically replicate itself in an unstructured environment. Our analysis showed that, in order to start the assembly process of the new replica, the robot has to move around, find the parts, pick them up and put them in place. For the robot's first challenge, we focused on the computer vision. The robot had to find, locate and track parts in the environment. For this

---

[4]An open source computer vision and machine learning software library (OpenCV 2017).

Fig. 5 First version of LEGO Mindstorms EV3 robot

challenge, we implemented a camera, but other combination of sensors such as sonar or infrared could also have been used to detect objects. We found that automated factories, construction and space development were motivating application scenarios for self-replicating robots. The robot performed well in recognizing single-colored objects in different colors with the implemented object recognition algorithm (see Fig. 6). However, the identification of different shapes using only a low-cost camera seemed to be more difficult. When an object was close to the camera we used the SURF algorithm (*Speeded up robust features*) (Bay et al. 2008) to identify whether the object was the exact one we were looking for. Based on the SIFT algorithm (*Scale-invariant feature transform*) (Lowe 2004), SURF was scale and rotation invariant, but faster and more robust compared to SIFT. SURF selected interest points of an image and then built local features based on histograms of gradient like local operators. However, the SURF algorithm could not differentiate between particular LEGO bricks, because SURF uses a feature-matching procedure and some bricks have too few distinguishing features for this algorithm. For future research, we could either aim to further improve the SURF algorithm so that it is usable in real-time systems, or pursue an alternative solution to recognize objects with an increased number of features.

Choosing the second scenario, in the next study (Hunkeler 2017), we implemented a feature-matching algorithm to the LEGO Mindstorms EV3 prototype, similar to Mino (2015) which is capable of remembering and identifying specific items. As an alternative, we employed ORB algorithm (*Oriented FAST and rotated BRIEF*) (Rublee et al. 2011), which can outperform SIFT and SURF in certain situations. The developed object recognition robot is capable of taking pictures of an object from multiple viewpoints using feature matching. The vision behind this feature-matching is to identify not any LEGO brick, but rather, exactly the LEGO brick that

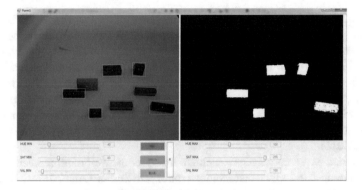

Fig. 6 Multiple-object recognition based on color awareness

the robot is looking for, hence bringing our prototype one small step closer to the self-replication ideal.

We managed to increase recognition precision significantly when comparing our four-viewpoint algorithm to a single image version of the same algorithm. However, we were not able to satisfactorily confirm performance with the evaluation data set. Our analysis showed that the proposed approach only worked well in combination with images that have strong distinctive features, such as text. If such features are missing, the algorithm produces many false positive results and thus delivers an overall unacceptably low accuracy rate. Currently, our approach is to use four different viewpoints. In the future, we plan to analyze whether a different number of viewpoints could yield better results.

3.2 Autonomous Navigation Towards Self-Replicating Robots

Object recognition poses a challenge for self-replicating robots as does navigation, in particular, obtaining accurate location information without any dependency on further infrastructure. This enables independent and broad usage of robots without any additional effort. In our research (Meier 2016), we presented the feasibility of autonomous indoor navigation with a developer's edition tablet from Google called Project Tango (Google 2016). We summarized that autonomous navigation requires three fundamental components in order to work: (1) The robot must be location-aware, (2) the robot must be able to sense its surroundings to gain knowledge about its environment, and (3) the robot needs AI to recognize and interpret the information to derive control commands using such data. With the exception of the AI component, Project Tango includes all the necessary features such as motion tracking, area learning and depth perception. The quality, accuracy and reliability of the Project Tango system was therefore assessed and discussed. It was proven that the accuracy of the motion tracking component is only high and stable if the tests are conducted under laboratory conditions. For ad-hoc applications, the results fluctuated between sufficient and poor. However, we found the depth perception capability to be quite accurate within the official operation range, although the quality was not sufficient in terms of the resolution of the measurement. Overall, the evaluated release of Google's Project Tango was relatively unstable and not very reliable for every serious, productive application. The evaluated tablet is no longer available on the market. However, in 2016, Lenovo launched the first consumer-ready smartphone with Tango technology and we have yet to explore its business application potential (Google 2016).

Therefore, we extended our research to find out how well reinforcement learning (RL) technique can apply to autonomous robot path finding (Anh 2017). RL is based on the ideal that an agent can learn a transition model for its own moves and can perhaps learn to predict its opponent's moves. The agent needs feedback about what is good and what is bad in order to decide what it has to learn (Russell and Norvig 1995; Mnih et al. 2013). We challenged the LEGO Mindstorms EV3 robot with RL on a grid map environment and made a comparison between Q-learning (Watkins 1989) and Deep Q-Network (Mnih and Kavukcuoglu 2015) as a self-learning agent. The most important contribution of this research is applied reinforcement learning and deep reinforcement learning as the robot's path-finding algorithm and the comparison between them. The simulation results show that both algorithms work well in the static environment. However, Deep Q-Network can perform at a better level of performance and stability than Q-learning in dynamic environment with other moving objects.

3.3   Further Research Topics in Self-Replicating Robots

Our main research steps in the visionary project on self-replicating, artificial intelligence robots have been to reduce the complexity step-by-step and tackle each research field in its unique environment: e.g. computer vision, movement in the sense of path finding in dynamic environments, and learning. We are gaining knowledge of algorithms and behaviors and conducting experiments to understand and implement tougher challenges for robots. When imagining a business use case, such as service robots, these robots will not gather LEGO bricks in real-life scenarios, but with the gathered knowledge we can assign them complex tasks such as *"find my keys!"* and send them on an expedition to retrieve a lost object. In order to comprehend such tasks, robots will then render all data input provided by the environment and run complex algorithms to establish a solid and reliable operational value.

4 Conclusions and Outlook

In this chapter, we have discussed our prototype-based research philosophy of building prototypes by using selected gadgets investigating certain (business) cases. We have presented some of our research projects in the two fields of immersive virtual reality and self-replicating robots to emphasize our research philosophy. In VR, the gadgets are used to identify new business applications by extending VR with gesture and voice control and allowing immersive 3D data analytics. In addition, a medical application of claustrophobia treatment was discussed. In the field of self-replicating robots, we used model robots and embedded artificial intelligence algorithms to let them solve certain sub-tasks related to the broad vision of self-replicating robots. We learned to understand and ask the right questions towards the development of self-replicating robots.

The development of prototypes and use cases, such as the previously mentioned VR applications or LEGO Mindstorms prototypes, and their continuous improvement and transition to real-world applications are painstaking processes. As described in these examples, our research philosophy pursues the idea of constantly challenging new technologies on a model level by setting more complex tasks, posing more limitations and approaching them with more creativity. With this vision, we keep building prototypes addressing these challenges and inventing possible solutions for real-world scenarios in such limited settings. For instance, a VR headset might not yet be able to fulfill all the necessary requirements to bring the industry a step forward (e.g. the voice-steered navigation). However, with innovation awareness and creative processes, we have no doubt that the next generation of gadgets will provide such necessary improvements. Our research philosophy aims to find out (1) how to solve real-world problems by modeling the real world, developing prototypes using gadgets, finding solutions on the model level and transferring them back to the actual use cases, (2) how to develop business applications, and/or (3) how to increase the efficiency of gadgets at the same time.

On this account, last, but most important is that in the middle of the last century Heidegger (1954, p. 100) said "the essence of technology is by no means anything technological". The acceptance of any new gadget is not about its technological capabilities or solely about its business readiness but also about how it is integrated in our social, cultural and behavioral environments, as remarked in Davis's "*Technology Acceptance Model*" (Davis 1989; Venkatesh and Davis 2000). Looking back, society has experienced technological advances through smartphones and smart glasses within the communication sector. Nowadays, we are experiencing the paradigm shift in the collaboration and automation sectors with the leap provided by "*Internet of Things*" trends (Gubbi et al. 2013) and with the introduction of AR applications with a strong business application focus (Etherington 2017). In this context, the Industry 4.0 ecosystem provides a solid and unified approach to this paradigm shift in technology acceptance by using smarter interactions and mediators (i.e. gadgets) to gather necessary data and make meaningful and wise predictions out of it. Until the market delivers such smart gadgets with immediate learning and knowledge management to the consumers, our research projects will focus on either combining different resources (gadgets with other gadgets/services) or embedding new algorithms with the goal of getting closer to this paradigm. Bearing in mind these reflections, in order to initiate technological innovation and for a better adoption of these tools, we are also steering our research towards the diffusion of these technologies and applications in a broader context and a variety of business opportunities.

References

Altarteer S, Vassilis C, Harrison D, Chan W (2016) Product customisation: Virtual reality and new opportunities for luxury brands online trading. In: Proceedings of the 21st international conference on Web3D technology, ACM, New York, NY, USA, Web3D'16, pp 173–174. https://doi.acm.org/10.1145/2945292.2945317

Anh HT (2017) An autonomous path finding strategy for an artificial intelligence enabled lego mindstorms robot. University of Applied Sciences and Arts Northwestern Switzerland. MSc in Business Information Systems, dual-degree Master Thesis

Barnes S (2016) Understanding virtual reality in marketing: nature, implications and potential

Bay H, Ess A, Tuytelaars T, Van Gool L (2008) Speeded-up robust features (surf). Comput Vis Image Underst 110(3):346–359

Bhuiyan J (2017) Alphabet's self-driving cars drove their last million miles in record time—recode. https://www.recode.net/2017/5/10/15605054/alphabet-waymo-self-driving-3-million-miles. Accessed on 13 July 2017

Bigné E, Llinares C, Torrecilla C (2016) Elapsed time on first buying triggers brand choices within a category: a virtual reality-based study. J Bus Res 69(4):1423–1427

Brennenstuhl M (2017) Using voice commands to enhance the navigation in virtual reality on a smart phone. University of Applied Sciences and Arts Northwestern Switzerland. MSc in Business Information Systems, Master thesis

Buff E (2015) Exploring and developing use cases for oculus rift combined with gesture control systems within the health industry. University of Applied Sciences and Arts Northwestern Switzerland. BSc in Information Systems, Bachelor thesis

Carlin AS, Hoffman HG, Weghorst S (1997) Virtual reality and tactile augmentation in the treatment of spider phobia: a case report. Behav Res Ther 35(2):153–158

Chirikjian GS, Zhou Y, Suthakorn J (2002) Self-replicating robots for lunar development. IEEE/ASME Trans Mech 7(4):462–472

Davis FD (1989) Perceived usefulness, perceived ease of use, and user acceptance of information technology. MIS Quart pp 319–340

Dede C (2009) Immersive interfaces for engagement and learning. Science 323(5910):66–69

Doganci F (2015) Wearable technology: From science-fiction to reality, analysis of head-mounted-displays in technology and science-fiction. University of Applied Sciences and Arts Northwestern Switzerland. BSc in Information Systems, project Report in Topics in Business Information Technology

Etherington D (2017) Google glass is back with hardware focused on the enterprise | techcrunch. https://techcrunch.com/2017/07/18/google-glass-is-back-with-hardware-focused-on-the-enterprise/. Accessed on 21 July 2017

Facebook (2017) Facebook spaces. https://www.facebook.com/spaces. Accessed on 13 July 2017

Fei-Fei L (2015) How we're teaching computers to understand pictures. https://www.ted.com/talks/fei_fei_li_how_we_re_teaching_computers_to_understand_pictures

Fei-Fei L, Fergus R, Perona P (2006) One-shot learning of object categories. IEEE Trans Pattern Anal Mach Intell 28:594–611

Frey G, Jurkschat A (2017) Intuitive hand gestures for the interaction with information visualizations in virtual reality. University of Applied Sciences and Arts Northwestern Switzerland. MSc in Business Information Systems, project Report in Innovative Thinking Project

Gabbatt A (2011) IBM computer Watson wins jeopardy clash. https://www.theguardian.com/technology/2011/feb/17/ibm-computer-watson-wins-jeopardy. Accessed on 29 June 2017

Gartner (2017) Top trends in the gartner hype cycle for emerging technologies, 2017—smarter with gartner. http://www.gartner.com/smarterwithgartner/top-trends-in-the-gartner-hype-cycle-for-emerging-technologies-2017/. Accessed on 07 Oct 2017

Gartner (n.d.) Hype cycle research methodology. http://www.gartner.com/technology/research/methodologies/hype-cycle.jsp. Accessed on 07 Oct 2017

Google (2016) Tango. https://get.google.com/tango/. Accessed on 14 July 2017

Google (2017) Google VR. https://vr.google.com/. Accessed on 17 July 2017

Gubbi J, Buyya R, Marusic S, Palaniswami M (2013) Internet of things (IoT): A vision, architectural elements, and future directions. Future Gener Comput Syst 29(7):1645–1660. https://doi.org/10.1016/j.future.2013.01.010, http://www.sciencedirect.com/science/article/pii/S0167739X13000241

Halsey E (2016) 5 ways virtual reality will change architecture. http://archsmarter.com/virtual-reality-architecture/, Accessed on 29 June 2017

Heidegger M (1954) The question concerning technology. Essential readings, Technology and values, pp 99–113

Hey J (2004) The data, information, knowledge, wisdom chain: the metaphorical link. Intergovernmental Oceanographic Commission 26

Hotho S, McGregor N (2013) Changing the rules of the game: economic, management and emerging issues in the computer games industry. Springer

Huang YC, Backman KF, Backman SJ, Chang LL (2016) Exploring the implications of virtual reality technology in tourism marketing: an integrated research framework. Int J Tourism Res 18(2):116–128

Hunkeler I (2017) Object recognition using multiple viewpoints. University of Applied Sciences and Arts Northwestern Switzerland. MSc in Business Information Systems, Master thesis

IBM Watson (2017) IBM Watson—build your cognitive business with IBM. https://www.ibm.com/watson/. Accessed on 14 July 2017

Lavanchy L (2015) Virtual reality for online stores—case of oculus rift. University of Applied Sciences and Arts Northwestern Switzerland. BSc in Information Systems, project Report in Topics in Business Information Technology

LeapMotion (2010) Leap motion. https://www.leapmotion.com/. Accessed on 20 June 2017

Lowe DG (2004) Distinctive image features from scale-invariant keypoints. Int J Comput Vis 60(2):91–110

Lui TW, Piccoli G, Ives B (2007) Marketing strategies in virtual worlds. SIGMIS Database 38(4):77–80. http://doi.acm.org/10.1145/1314234.1314248

Meier DM (2016) Google project tango enables the lego mindstorms EV3 robot model to perform autonomous indoor navigation. University of Applied Sciences and Arts Northwestern Switzerland. MSc in Business Information Systems, Master thesis

Merriam-Webster (n.d.) Gadget | definition of gadget by merriam-webster. https://www.merriam-webster.com/dictionary/gadget. Accessed on 23 July 2017

Milgram P, Kishino F (1994) A taxonomy of mixed reality visual displays. http://etclab.mie.utoronto.ca/people/paul_dir/IEICE94/ieice.html. Accessed on 11 July 2017

Mino K (2015) Analysis and application of computer vision principles towards self-replicating robots. University of Applied Sciences and Arts Northwestern Switzerland. MSc in Business Information Systems, dual-degree Master thesis

Mnih V, Kavukcuoglu K (2015) Methods and apparatus for reinforcement learning. https://www.google.com/patents/US20150100530, US Patent App. 14/097,862

Mnih V, Kavukcuoglu K, Silver D, Graves A, Antonoglou I, Wierstra D, Riedmiller M (2013) Playing Atari with deep reinforcement learning. arXiv:13125602

Moses MS, Ma H, Wolfe KC, Chirikjian GS (2014) An architecture for universal construction via modular robotic components. Robot Auton Syst 62(7):945–965

Napp N, Burden S, Klavins E (2006) The statistical dynamics of programmed self-assembly. In: Proceedings 2006 IEEE international conference on robotics and automation, ICRA'2006. IEEE, pp 1469–1476

OculusRift (2015) Oculus rift development kit 2 (dk2) | oculus. https://www3.oculus.com/en-us/dk2/. Accessed on 12 July 2017

OpenCV (2017) About—openCV library. http://opencv.org/about.html. Accessed on 14 July 2017

Rublee E, Rabaud V, Konolige K, Bradski G (2011) Orb: an efficient alternative to sift or surf. In: 2011 IEEE international conference on computer vision (ICCV). IEEE, pp 2564–2571

Russell J (2017) Google's AlphaGo Ai wins three-match series against the world's best go player | techcrunch. https://techcrunch.com/2017/05/24/alphago-beats-planets-best-human-go-player-ke-jie/. Accessed on 13 July 2017

Russell S, Norvig P (1995) Artificial intelligence: a modern approach (3rd edn). https://www.amazon.com/dp/0136042597/ref=cm_sw_r_cp_ep_dp_GkDozbZQXY8SV. Accessed on 17 July 2017

Safrudin N, Fay M, Changa A, De Wit B (2015) Game changing digital technologies. https://www.researchgate.net/publication/308694636_Game_Changing_Digital_Technologies. Accessed on 29 June 2017

Schär F (2017) Using virtual reality to enhance exploratory analysis of categorized financial data. University of Applied Sciences and Arts Northwestern Switzerland. MSc in Business Information Systems, Master thesis

Stansfield SA, Miner N, Cooke C (2016) Exploring the application of virtual reality to remote robot operations. Int J Virtual Reality (IJVR) 02(2):1–11. https://hal.archives-ouvertes.fr/hal-01530723

Stone LM, Erickson T, Bederson BB, Rothman P, Muzzy R (1994) Visualizing data: is virtual reality the key? In: 1994 Proceedings of the IEEE conference on visualization, Visualization'94. pp 410–413. https://doi.org/10.1109/VISUAL.1994.346286

Suthakorn J, Cushing AB, Chirikjian GS (2003) An autonomous self-replicating robotic system. In: 2003 Proceedings of the IEEE/ASME international conference on advanced intelligent mechatronics, AIM'2003, vol 1. IEEE, pp 137–142

Vakulenko M (2013) Asymmetric business models and the true value of innovation. https://www.slashdata.co/blog/2013/02/asymmetric-business-models-and-the-true-value-of-innovation. Accessed on 17 July 2017

Venkatesh V, Davis FD (2000) A theoretical extension of the technology acceptance model: four longitudinal field studies. Manage Sci 46(2):186–204

Wang H, Tian F, Gao B, Bian J, Liu TY (2015) Solving verbal comprehension questions in IQ test by knowledge-powered word embedding. ArXiv e-prints

Watkins CJCH (1989) Learning from delayed rewards. PhD thesis, King's College, Cambridge

Watts T, Swann G, Pandit NR (1998) Virtual reality and innovation potential. Bus Strateg Rev 9(3):45–54

Weise S, Mshar A (2016) Virtual reality and the banking experience. J Digital Bank 1(2):146–152. http://www.ingentaconnect.com/content/hsp/jdb001/2016/00000001/00000002/art00007

Wolfram S (2015) Wolfram language artificial intelligence: the image identification project. http://blog.stephenwolfram.com/2015/05/wolfram-language-artificial-intelligence-the-image-identification-project/

Co-robots from an Ethical Perspective

Oliver Bendel

Abstract Cooperation and collaboration robots work hand in hand with their human colleagues. This contribution focuses on the use of these robots in production. The co-robots (to use this umbrella term) are defined and classified, and application areas, examples of applications and product examples are mentioned. Against this background, a discussion on moral issues follows, both from the perspective of information and technology ethics and business ethics. Central concepts of these fields of applied ethics are referred to and transferred to the areas of application. In moral terms, the use of cooperation and collaboration robots involves both opportunities and risks. Co-robots can support workers and save them from strains and injuries, but can also displace them in certain activities or make them dependent. Machine ethics is included at the margin; it addresses whether and how to improve the decisions and actions of (partially) autonomous systems with respect to morality. Cooperation and collaboration robots are a new and interesting subject for it.

Keywords Cooperation robots · Collaboration robots · Co-robots
Cobots · Information ethics · Technology ethics · Business ethics
Machine ethics

1 Introduction

In the factories of old, robots were mostly housed in protected areas. New factories offer a different sight. Transport robots roll back and forward in the halls with their loads, transporting parts and goods from one place to another. If a human crosses their path, they brake and wait until he or she is at a safe distance. Nowadays, these robots are mostly underway on virtual tracks, but in the future, where reasonable or necessary, they will be given greater freedom. Cooperation and collaboration

O. Bendel (✉)
Institute for Information Systems, University of Applied Sciences and Arts Northwestern
Switzerland, Bahnhofstrasse 6, 5210 Windisch, Switzerland
e-mail: oliver.bendel@fhnw.ch

© Springer International Publishing AG 2018
R. Dornberger (ed.), *Business Information Systems and Technology 4.0*,
Studies in Systems, Decision and Control 141,
https://doi.org/10.1007/978-3-319-74322-6_18

robots work in close proximity to and hand in hand with human colleagues. They take on cumbersome activities from humans, which have worn out their joints and overexerted their muscles. Apparently there is also a paradigm shift in the context of Industry 4.0 because the cooperation (and no longer the separation) of machine and man turns out to be the silver bullet to success, at least for the time being.

This article deals with cooperation and collaboration robots. They are defined and classified, and areas of use, application and product examples are mentioned. Against this background, a discussion on moral issues takes place, from the perspectives of information and technology ethics as well as of business ethics. Key terms from these disciplines are applied to the areas of application. Machine ethics as a design discipline also comes into play briefly. Not least, the robot tax is mentioned because in the case of cooperation and collaboration robots it can be demonstrated that it is difficult to implement this tax in practice, even if one may advocate it in theory. This brings us back to information and business ethics, and to the moral and societal implications of the use of autonomous machines. The conclusion consists of a summary and an outlook.

2 Terms and Characteristics of Co-robots

In order to approach the much-used and little-explained concepts of cooperation and collaboration robots, it is advisable to look at the components of the composites in more detail. Cooperation cannot be equated with collaboration, although the necessary care is not always taken when using the terms on their own or in composites, and not every machine, or even every automat is a robot. So, what is a robot, and how and with whom does it cooperate and collaborate where necessary?

Cooperation means teamwork, in particular in the scientific, political or economic fields (Bendel 2017, p. 4). Typical is a high degree of division of labor and the simultaneous existence of a common goal. With collaboration, a high level of cooperation is assumed. The activities are intertwined and build on one another. In both phenomena, a form of cooperation is thus given, with one being the common goal, and the other the common task (and the common path). In this respect, their dependence in time and space has to be assessed differently. In cooperation, this is not as high as in collaboration. With appropriate agreements, the partners can perform their tasks relatively independently, in different locations and at different times. A mandatory submission or finish date is often given in both, in industry almost without exception, because cooperation of course is not an end in itself, but an (efficient and effective) means to an (economic) purpose.

Christaller defines robots as sensomotoric machines for the enhancement of the human capacity for action (Christaller et al. 2001, p. 5). Accordingly, they would consist of mechatronic components, sensors, and computer-based monitoring and control functions. The complexity of a robot differs significantly from other machines by the greater number of degrees of freedom and the variety and scope of its behavior (Christaller et al. 2001, p. 5). The degree of freedom is defined through axes and joints. In addition to extending the capacity to act and the abolition of the possibility to work, the partial or complete substitution of human beings by the machine could be mentioned. The ability to make decisions becomes increasingly relevant, and human autonomy (from "autós", "self", and "nómos", "law", literally "self-law"), which in turn allows one to think of freedom (also of foreign law), is supplanted by the machine. It is possible to divide robots into different types such as industrial robots, service robots, space robots and combat robots, and also into hardware robots—which these types can be categorized as—and software robots such as chatbots, social bots, agents and crawlers. For several years now, a bridge has been built between industrial and service robots, and it can be said that cooperation and collaboration robots contain traces of the second mentioned type, or that they are usually industrial robots, but can also act as service robots, for example in the care sector.

The intensive preoccupation with cooperative and collaborative robots began in the 1990s (Peshkin 1996). They may also be called "cooperation or collaboration robots" or, abbreviating, generalizing, and following the English model, "co-robots" and "cobots" (in the German variant, "Kobots"). As implied, the differences in meaning can be taken more or less seriously. It is all about cooperating and collaborating with humans (Bendel 2017, p. 4). This may also take place among machines, but there it is common and widespread at a certain level, whereas close cooperation between industrial robots and employees, for example, is novel and mostly profitable for both sides. Cooperation robots work with people step by step for a common goal; collaboration robots, on the other hand, work with people hand in hand on a common task (with a specific goal). They use their mechanical and sensory skills to make decisions with regard to products and processes.

Cooperation and collaboration robots usually have one arm or a pair of arms and two to three fingers. Several axes or joints (degrees of freedom) allow appropriate mobility and adaptability. In a certain sense and scale, they are designed to complement humans, who also have arms and fingers, and both can approach the product in a similar way and "comprehend" it. The majority are lightweight robots that can be moved between places, i.e. they are mobile at least in this passive sense (Bendel 2017, p. 4). They cooperate or collaborate with people; they get very close to them and their activities can mesh. Despite this close cooperation, a high level of safety is expected in the workplace, especially with respect to humans, who must not be injured, but rather protected and relieved. These robots are relatively slow, at least in the shared space, have quite natural movements, and in the best case can distinguish between objects and humans, in order to function appropriately with each.

Co-robots are autonomous, intelligent, adaptive systems and are created to be generalists, whereby the changes on the software side must have their equivalents on the hardware side, for example in the way tools and gripping hands can be replaced and extended. In a given production situation, it is of particular interest that they can learn from people moving their arms or showing them something in front of the cameras, and permitting machine learning and mechanical or humans to form a beneficial alliance. In this regard, it is important that robots are designed to be similar to humans in certain aspects, because, for example, they can only imitate movements of the limbs if they have them too.

3 Examples of Cooperation and Collaboration Robots

Having established the basics, some examples of co-robots will be presented below. Almost every self-respecting industrially oriented robotics company has a co-robot in its program. There are common features, but also differences, not only in design and function, but also in the approach to marketing. The following descriptions use "cooperative" and "collaborative" as far as possible in accordance with the wording of the manufacturers (Bendel 2017, p. 4 ff.):

- Baxter by Rethink Robotics was completed in 2012 and launched for the first time in 2013. It is a cooperative robot with three cameras and a 360° sonar, with which it detects obstacles. In addition, it has a display in the form of a tablet. Baxter has two arms, each with seven joints or degrees of freedom. It can be equipped with an electric gripper of variable size or with a pneumatic gripper. The company emphasizes that it is not necessary for the user to program Baxter, but to train it, for example by moving one's arms. This cooperative robot can take on different tasks in production.
- The collaborative robot Sawyer from the same company also has seven degrees of freedom, although with only one arm. Sawyer is smaller and lighter than its predecessor, Baxter. Worthy of mention are the flexible motion control and the integrated imaging system. It can pick up, reposition, and correctly place objects that are not exactly aligned, similar to the function of graphics programs. It also has a display. Sawyer is ideal, for example, for machine assembly and printed circuit board tests, as shown on the website www.rethink-robotics.com.
- YuMi (according to the website www.abb.com, "Yu" stands for "You" and "Mi" for "Me", which emphasizes working in tandem) by ABB, presented at the Automatica 2014, is a collaborative, two-armed robot for small part assembly. It has flexible, skillfully gripping hands, parts feeding systems, and a camera-based part detection. The range of the robotic arms corresponds roughly to the range of human limbs. According to the manufacturer, in the event of an impending collision with a human colleague, it can stop its movement within milliseconds.
- The lightweight robot LBR iiwa, presented in 2013 at the Hannover Messe, is referred to by Kuka (www.kuka.com) in different announcements as both cooperative and collaborative. LBR iiwa has an arm with seven axes. It can be taught movement patterns by demonstration. The company describes this as simulation on its website. The LBR iiwa is an example of what is referred to as a sensitive robot and is used in the assembly of dishwashers, where adjustments to new models can be made relatively easily.
- Franka Emika by KBee, presented at the Hannover Messe in 2016, had a seven-axed, torque-based, also sensitive robotic arm. With a few clicks, workflows could be compiled in a graphical user interface. What was special about Franka Emika, however, was its price of about 10,000 €. This means that SMEs, even private households, could flirt with the idea of buying such a robot. The founder of KBee later established Franka Emika (www.franka.de). This company produces the co-robot Panda.

- Jaco[2] 6 DOF, commonly known as Jaco, is a product of Kinova Robotics (www. kinovarobotics.com). It is an arm with a hand possessing three fingers, which can help people with limited arm and hand functions by handing them something or opening something for them. It moves in six degrees of freedom. Jaco's example proves that co-robots have their place and are able to perform their services not only in the factory, but also in private households and care homes—as assistants and care robots.

Most co-robots have good proportions, round shapes and muted colors (Bendel 2017, p. 5). Their surface is smooth or slightly roughened. Their external design alone should make them appear attractive and not at all frightening. They are usually not humanoid, in the narrow sense of the term (because a pair of arms can seem quite human, even if a body and legs are missing). In the cases of Baxter and Sawyer, the display can be used to represent a face. Mimicking abilities can be quite advantageous for co-robots, but involve the danger of distraction; workers should be looking primarily at their hands and the hands of the robot. In addition, they must pay attention to harmonious interaction in order not to jeopardize the intended communication improvement. The reason for the imitation of limbs and hands, as mentioned before, is due to the requirements of cooperation and collaboration. In addition, these are generally tried and tested forms, and if the robot's arms are not too thick and have sufficient degrees of freedom, it can both work in a space-saving way and be stowed away easily. This in turn is of importance for some factories as well as for care and therapy facilities or for private households.

4 Cooperation and Collaboration Robots in Industry 4.0

This chapter describes the role of cooperation and collaboration robots in Industry 4.0. The systematization essentially follows Bendel (2015a). The fourth industrial revolution is based on the third, the revolution of digitalization. One could just as well say that the third revolution is experiencing a change of direction, in order to avoid the inflationary creation of terms and numbers. Industry 4.0 is characterized by automation, autonomization, flexibilization and individualization in digitalization, with a networking as complete as possible, and by increasing efficiency and effectiveness as much as possible (Bendel 2015a, p. 740). The focus is on the smart factory working busily and cleverly, the intelligent, supposedly anticipatory, judging and decisive factory. It operates with the help of cyber-physical systems (which consist of physical components, receive virtual inputs and produce physical products) and innovative industrial robots, and is connected to its environment (Bauernhansel 2014; Bendel 2015a, p. 740). Industry 4.0 is considered and explained by the disciplines of business administration and information systems as well as computer science (Herda 2014), where not only core computer science, but also artificial intelligence and human-computer interaction play a role. Also of relevance is, of course, robotics. In the following description, we will refer to these concepts and establish the connection with co-robots:

- Automation has a long tradition that goes back to ancient times. Industry 4.0 is about electronically-controlled automated production, automated production facilities and automated data transmissions that in turn influence production. In cooperation and collaboration robots, automation is special, because it is important to respect the workers as the acting subjects and the objects to be treated, who are not subordinate to it. Defined procedures should be in place, and it should also be possible to take unforeseen events such as technical or human failure into consideration. In addition, changes of staff with varying degrees of education and experience have to be absorbed.
- The concept of autonomy refers to the independence of people and machines. In Industry 4.0, the one increasingly replaces the other. In the course of this development, which ends in the autonomization of the machines, humans are of secondary importance. They control and maintain the machines that make decisions and begin to act. With cooperation and collaboration robots, autonomization can be driven forward, but can also be rejected. On the one hand, the machine needs to make further autonomous decisions, especially in cooperation with humans, for example, in order not to injure, but rather to support them. On the other hand, however, humans in the process can be allowed to prevent and overrule certain decisions made by the machine.
- Flexibilization is mentioned when there is a prompt and appropriate response to requirements. Production in the intelligent factory is accelerated, slowed down, stopped, realigned and arranged from one minute to the next. Other items are mass produced or special ones are produced within the series, created for example with the help of 3D printers. In the context of cooperation and collaboration robots, flexibilization is gaining new meaning. The input comes not only from databases, information systems and social media, but also directly from the human counterpart as part of the process. In this way, internal and external sources of information can be relativized and spontaneous decisions taken, which can ultimately come in the form of instructions from leaders and supervisors.
- Individualization usually depends on wishes, which relate to form, function and content and which are based on the interested person or customer. These are involved via participatory media so that "individualization" can affect the individual or group, and via other digital and traditional channels. The result are hybrid (material and virtual) products that match the customer's wishes exactly not only in their nature and quality, but also in terms of service and insurance benefits. The consumer will in some cases become a producer, a prosumer, to be more specific. Individualization is often related to flexibilization. In cooperation and collaboration robots, the specialist can have specific influence. He or she can ensure that individualization is implemented even more in the interest of the customer, due to the fact that she or he has a share in the individualization, for example by means of personal interventions related to the product.

Networking in Industry 4.0 therefore includes things, systems and people, specifically plant managers, employees, workers, customers, etc. in value creation processes, bringing together internal staff and external customers, people and machines. A close relationship exists between cooperation and collaboration robots and specialist or assistant forces. The distinctive feature seems to be that networking here is carried out between technical and non-technical resources and between automated and individual processes (which is, of course, known from the partially automated production). However, the human who works with the machine can be engineered via chips, data glasses, exoskeletons, etc., and thus become a cybernetic organism, known as a cyborg. This phenomenon will be considered below from an ethical perspective.

5 Information, Technology, Business and Machine Ethics

Following the fundamental clarification above of what ethics is and what it can do, the fields of applied ethics are outlined below, considered with regard to cooperation and collaboration robots. They have developed different terms and priorities in their methods. In addition, machine ethics is addressed, which as a design discipline does not only reflect on the use of robots, but also influences their development.

Ethics is a discipline of philosophy. One of its founders is Aristotle. It applies specific scientific methods for the purposes of justification and description. Its object is morality, and it is also known as moral philosophy, not least to distinguish it from moral theology, which has the same object, but completely different methods of examination. It cannot be regarded as scientific. Moral economics is also devoted to morality, but concentrates on the field that is undergoing moral scrutiny, namely the economy.

The object of information ethics is the morality of those who offer and use information and communication technology (ICT), application systems and new media (Bendel 2012b; Kuhlen 2004). It inquires how these persons, groups and organizations behave with regard to aspects of morality, and how they should behave. Those who do not offer and use ICT or new media but are involved in their production or are affected by their effects are also relevant. Thus, information ethics focuses on morality in the information society and analyses how its members behave, or should behave, in moral terms. It also analyses the relationships of the information society within itself, with non-technology affine members, and with low-tech cultures from the perspective of ethical aspects.

Technology ethics relates to moral issues in the use of technology. It can focus on the technology of buildings, vehicles or arms as well as on nanotechnology. In the information society, technology ethics is also closely connected to information ethics. Not least, it has to cooperate with business ethics, in as far as companies are involved in the development, production and marketing of technological products, and these are demanded and used by their customers.

The subject of business ethics is the morality of the economy. The focus is on the person who has economic interests, who produces, trades, leads and executes (various forms of individual ethics) as well as consumes (consumer ethics), and on the company that bears responsibility towards its employees, customers and the environment (corporate ethics). In the 21st century, business ethics is closely linked to information and technology ethics. It is either driven by ethics or, as mentioned already, by the economics, as moral economics. There are also integrative approaches.

Machine ethics is not only a reflection discipline, but also a design discipline, insofar as machines are conceived and constructed (Bendel 2012a; Bendel 2014a). The concept of robot ethics can be understood either as an aspect of machine ethics, with a focus on (the design of) robots, or as a reflection discipline like information and technology ethics, again with a focus on robots.

6   Moral Implications of Co-robots

The moral implications of the use of cooperation and collaboration robots are further distinguished by information, technology and business ethics, as described above. Another section addresses machine ethics. The focus is on Industry 4.0. The remarks are based on Bendel (2015a) and develop the findings of this source.

6.1   Information Ethics

In the new factory, the factory of Industry 4.0, automation is implemented by means of digitalization, the spectrum of which is fully exploited. Cooperation and collaboration robots are controlled with the help of computer technologies and the expertise of the present and higher-ranking employees. Information ethics requires reliability in this respect, for the safety of people with regard to programming, implementation and operation, as well as for the physical integrity of the cooperation and the mental and physical impairment caused by the use of personal data. This is automatically collected, processed and forwarded and then, under certain circumstances, judged by humans. Thus, liability is of interest, less in the legal than in the moral sense.

Autonomization is also associated with digitalization and with independent mechanical decisions, which involve different levels. Information ethics examines the consequences of decisions for humans. Here, too, security is required, but in a particular sense: For example, to what extent are the decisions of the machines good for the employee? Can she or he foresee and understand them? Whether and how can he or she cancel them if there is danger to life and limb? Furthermore, it is about taking part in semi-autonomous or autonomous decisions, which concern customers and consumers, prosumers and, in general, members of the information society.

Individualization can bring about losses in the informational autonomy. The customer's personal data, which are generated on a massive scale in Industry 4.0, are used in production, marketing, and distribution and may be passed on. In addition, workers at the side of the cooperation and collaboration robots may be able to access and misuse the data directly or by means of devices. They themselves can also be at risk in their informational autonomy: The robots might recognize, evaluate and monitor human behavior with their sensors and algorithms. This results in data being produced at close range and from problematic perspectives.

The hacking of factory systems and their environment should be considered above all in the context of networking. This tends to result in greater susceptibility to the extent that more vulnerabilities and gateways are present. Hacking can lead to the takeover of the systems (Kagermann et al. 2013, p. 51). In cooperation and collaboration robots, a takeover can result in risks and injuries to the workers. In addition, an operational failure in a machine can jeopardize the human's ability to work. The unauthorized intrusion into the systems can in turn affect informational autonomy and data protection, and it is possible for attackers to scrutinize the target objects, even for a longer period.

6.2 Technology Ethics

Automation always means mechanization. In the information age, this leads to an evident omnipresence: Technology dominates the factory halls, reduces the interactions between humans or replaces them entirely. In the case of co-robots, however, it is ensured, at least for the time being, that people are not being made completely redundant. Automation also means increasing dependence on technology. Those who are still present in the factory depend on it in every step of the way, which is evident in cooperation and collaboration robots. The human-follows-machine principle, which was also created in the context of information ethics (Bendel 2012b), is relevant. The concepts of cooperation and collaboration must not belie the fact that there is often a dominance of the machine and that humans must try to understand not only the logic, but also the "epic" of the machines, the way they present themselves, and how they develop.

Autonomization converts the omnipresence of technology into an omnipotence. The decisions and actions of cooperation and collaboration robots can have a direct impact on the employee. It is precisely his or her task—and again in the sense of the man-follows-machine principle—to direct the behavior of the robot towards certain approaches. It learns via its cameras and sensors by imitation of the human, or by the fact that its arms and hands are moved and rotated by simulation. In this respect, human autonomy is transferred to mechanical autonomy, and an image of decisions and actions is generated, in which it is the task of the machine to compensate for inaccuracies and irregularities and to correct errors.

As mentioned before, networking enables the hacking of electronic components and technical systems. As a result, someone can penetrate the systems and take over information technology components and also manipulate and ruin the technology, thereby altering the products or making production impossible, which poses a challenge to the human being. He or she monitors the technology for her or his own benefit, being continually at its mercy, together with the IT and corporate executives. As discussed in the last chapter, technology takes on a surveillance function.

6.3 Business Ethics

Automation often implies that human labor is replaced (Hirsch-Kreinsen 2014, p. 18). Those affected are mostly simple workers, whose manipulations and finger skills can be imitated by machines, while IT specialists and managers as their inventors and sponsors remain indispensable, at least in the early phases of the revolution. However, automation also implies that machines and humans cooperate in work cells, with the desired side effect of physical relief and a higher level of safety. In the use of cooperation and collaboration robots, the presence of humans is even a prerequisite and an integral part of the processes, which can, of course, be called into question again at a later date.

Autonomization in the context of business ethics is linked to the question of whether the business and economic consequences are positive or negative. The smart factory can make autonomous decisions which may endanger its existence, its medium and long-term production, the jobs that depend on it, and its location. At the same time, the smart factory is able to gain competitive advantage through decisions that have been made and implemented quickly. In cooperation and collaboration robots, the question arises regarding to what extent the machine's autonomy will displace that of humans and how one can and should define oneself as a worker. The robot is no longer a mere tool; instead it uses humankind as a tool.

Flexibilization is achieved by autonomous machines, which, in the case of co-robots, operate an immediate exchange with other systems of adequate data and information, and receive input via sensors. A fast connection and close networking are necessary, as well as access to and appropriation of high quality and up-to-date information in order to be able to assess the market environment and development. In addition, the use of specific robots requires specialized workers or engineered workers who in their function as cyborgs can meet the flexibility requirements. This again raises the question of the worker's identity, and whether the work is suitable for humans. There is a difference between wanting to be a cyborg and having to be one. A temporary, but even more, a permanent human enhancement, seen as a technical extension and improvement of the human, is a profound intervention in personal autonomy, and a transhumanism (an international movement that promotes the self-determined advancement of humans by means of scientific and technical resources) which is economically and not intrinsically motivated, is likely to provoke particularly broad rejection among those affected (Bendel 2015b).

6.4 Machine Ethics

Autonomization in automation involves machine decisions, which are also interconnected by means of networks, and have moral connotations, for example because they influence someone for good or ill. Machine ethics can try to teach moral skills to production facilities, stationary and mobile robots and individual devices such as 3D printers (Anderson and Anderson 2011; Bendel 2014b). Partially autonomous and fully autonomous systems can make the wrong decisions because they either obey inappropriate rules or interpret situations and operations incorrectly. They can injure people and cause accidents (Kagermann 2013, p. 51), which, in addition to machine ethics, social robotics attempts to counteract. Social robotics is mainly about socially compatible machines, for example to ensure that they react and act adequately, in working cells, on their way through the halls, in the areas of care and therapy, and above all, when they offer assistance in personal care and health matters.

In principle, cooperation and collaboration robots can be conceived as moral machines. However, they have not yet been the subject of machine ethics (Bendel 2014b), but rather, that of social robotics. Below, questions are asked from the perspective of machine ethics, without the possibility of giving answers as yet:

- Should a cooperation or collaboration robot be able to act morally?
- According to which individual or social morality should it act or according to which model of normative ethics (deontology, ethics of consequences, ethics of virtue, etc.)?
- Should it be able to refuse instructions from the worker, in the short, medium or long term, and if so, which instructions should it follow, and why?
- Should it in the first instance consider its human counterpart during cooperation or prioritize the product to be created, or the company or the market?
- Does there have to be a kill switch, as requested by a working group of the European Parliament, with which the robot can be turned off abruptly (Delvaux 2017), and if yes, who or what should be allowed to use it under what circumstances?
- Should it be able to ask self-improving questions, and should these follow moral requirements?
- Should it try to influence the motivation of the workers in a positive sense, and if so, should the person or rather the company benefit?
- Should it pay special attention to workers who are impaired or clumsy, or should it treat all partners at the workbench equally?
- Should it insist on learning only from skilled and experienced workers, and if so, will it be involved in their selection?
- Should it point out the inadequate performance of employees to them directly, or to their superiors?

Again, it becomes apparent that there are implications and consequences if the industrial robot leaves its incumbent place and gets close to humans, that it is, so to speak, crowding them, and if it works closely with them, becomes dependent and makes them dependent.

7 Robot Tax

The robot tax is an expression of the machine tax, which has been considered repeatedly since the 1950s and which can again be considered as a value-added levy (Bendel 2016; Becker n.d.). The idea is to tax the ownership or work of (partially) autonomous robots in production and other areas, and to give the funds either to the social security system or to the education sector. A link to the unconditional basic income is proposed repeatedly, although this can have quite different sources, for example a more consistent tax for companies.

In favor of a tax is the fact that the robot is recognized as a risk to full employment in the full-time model, and that a social and financial policy response to automation has to be found (Bendel 2016). Another positive fact is that there is talk about the economic and societal impact of robotics, due to automation and autonomization, culminating in the dangers of job losses, financial losses and loss of livelihoods. An argument against the tax is that the proliferation of robots that complement and relieve people is hindered, and that the proposed and intended liberation from the everyday burden of work could be prevented. Above all, however, it is not clear what exactly is to be taxed, which systems are affected and what kind of work is involved. This problem can be seen particularly well in this context. The activities of humans and machines are intertwined, and the question is whether the proportion of robot work is clearly determinable and whether a fictitious hourly wage to be taxed can easily be found.

Business ethics is needed to assess the implications for companies, employees and the unemployed, as well as corporate ethics and individual ethics. Business ethics can also be active at a macro level, in the sense of regulatory ethics, questioning the regulation of communities and countries. It is also possible to integrate technology ethics and information ethics, because they address the relationship between technology and humans and the use of information and communication technologies as well as the results of robotics and artificial intelligence.

8 Summary and Outlook

Cooperation and collaboration robots are changing the face of companies and operations, and are essential components of smart factories and even of Industry 4.0. They are penetrating other areas, in medicine and nursing, as tele robots for surgeries and as care robots that work in tandem with people, for example to move patients to different beds or lift them up, or to function as their extensions and facilities, if their radius of movement and action is restricted. In particular, investigation is needed regarding whether they really function as cooperation and collaboration robots in the true sense, or as other types of robots. From the perspective of ethics, there are many questions but the focus here is more on the risks. A further contribution is needed to address the opportunities. Above all, there is a preference for philosophical ethics in the unconditional sense and the openness of results, meaning that no fixed world or human image determines the image of technical and economic artifacts. The topic of the robot tax could be discussed on the sidelines, with respect to applied ethics.

For some, co-robots are the future of working, for others, an intermediate stage (Bendel 2017, p. 5). If they learn more and more, become more intelligent, more precise and even faster, humans could indeed become obsolete as partners and teachers. Then robots would work in tandem and in a team, and there would be interactions between machines, on an unprecedented level.

Robots would be able to help each other with particular tasks, as one robot would attach a tool to another and remove it again. Eventually, they would clean, repair and reproduce themselves. An inexhaustible field for machine-machine interaction is opening up. Maybe we will continue to be in demand for some time, if not in production, then at least in the household, in care homes and in hospital. There, we will continue to work side by side with co-robots, and if we are lucky, we will not only be of service to them, but they to us.

References

Anderson M, Anderson SL (eds) (2011) Machine Ethics. Cambridge University Press, Cambridge

Bauernhansl T, ten Hompel M, Vogel-Heuser B (2014) Industrie 4.0 in Produktion, Automatisierung und Logistik. Springer Vieweg, Wiesbaden

Becker J (n.d.) Maschinensteuer. Beitrag für das Gabler Wirtschaftslexikon. Springer Gabler, Wiesbaden (n.d.) http://wirtschaftslexikon.gabler.de/Definition/maschinensteuer.html. Accessed 9 Aug 2017

Bendel O (2012a) Maschinenethik. Gabler Wirtschaftslexikon. Springer Gabler, Wiesbaden. http://wirtschaftslexikon.gabler.de/Definition/maschinenethik.html. Accessed 9 Aug 2017

Bendel O (2012b) Informationsethik. Gabler Wirtschaftslexikon. Springer Gabler, Wiesbaden. http://wirtschaftslexikon.gabler.de/Definition/informationsethik.html. Accessed 9 Aug 2017

Bendel O (2014a) Wirtschaftliche und technische Implikationen der Maschinenethik. In: Die Betriebswirtschaft, 4/2014:237–248

Bendel O (2014b) Maschinenethik in der Industrie 4.0: Plädoyer für einfache moralische Maschinen. In: Wissenschaftsjahr 2014 – Die Digitale Gesellschaft, 12. Juni 2014. http://www.digital-ist.de/experten-blog/maschinenethik-in-der-industrie-40.html. Accessed 9 Aug 2017

Bendel O (2015a) Die Industrie 4.0 aus ethischer Sicht. In: Reinheimer S (ed) HMD – Praxis der Wirtschaftsinformatik, 52 (2015)5:739–748

Bendel O (2015b) Human Enhancement: Die informationstechnische Erweiterung und ihre Folgen. TATuP 2015(2):82–89

Bendel O (2016) Robotersteuer. Beitrag für das Gabler Wirtschaftslexikon. Springer Gabler, Wiesbaden 2016. http://wirtschaftslexikon.gabler.de/Definition/robotersteuer.html. Accessed 9 Aug 2017

Bendel O (2017) Co-Robots und Co. – Entwicklungen und Trends bei Industrierobotern. In: Netzwoche, 25 (2017)9:4–5

Christaller O et al. (2001) Robotik: Perspektiven für menschliches Handeln in der zukünftigen Gesellschaft. Berlin, Heidelberg, New York

Delvaux M (2017) REPORT with recommendations to the commission on civil law rules on robotics (2015/2103(INL)). A8-0005/2017, 24 Jan 2017. European Parliament, Brussels

Herda N, Ruf S (2014) Industrie 4.0 aus der Perspektive der Wirtschaftsinformatik. In: Wirtschaftsin-
    formatik & Management, 5/2014, pp 7–19
Hirsch-Kreinsen H (2014) Wandel von Produktionsarbeit - "Industrie 4.0". Soziologisches
    Arbeitspapier   Nr.   38/2014.   http://www.wiso.tu-dortmund.de/wiso/is/de/forschung/soz_
    arbeitspapiere/Arbeitspapier_Industrie_4_0.pdf. Accessed 9 Aug 2017
Kagermann H, Wahlster W, Helbig J (eds) (2013) Umsetzungsempfehlungen für das Zukunftspro-
    jekt Industrie 4.0: Abschlussbericht des Arbeitskreises Industrie 4.0. April 2013. http://www.
    bmbf.de/pubRD/Umsetzungsempfehlungen_Industrie4_0.pdf. Accessed 9 Aug 2017
Kuhlen R (2004) Informationsethik: Umgang mit Wissen und Informationen in elektronischen
    Räumen. UVK, Konstanz
Peshkin M (1996) "Cobots" work with people. In: IEEE robotics & automation magazine, vol 3(4),
    8–9, Dec 1996

Printed in the United States
By Bookmasters